# Better Queens

# by Jay Smith

Better Queens

Originally printed 1949 by Jay Smith

Reprint 2011 by

X-Star Publishing Company
Nehawka, Nebraska, USA
xstarpublishing.com

113 pages
40 Illustrations


ISBN 978-161476-051-1

# Transcriber's preface

I wanted this book available because I think Jay Smith was one of the great beekeepers and queen breeders of all time. There are many queen breeding books by scientists or small-scale breeders, but this is by a bee-keeper who raised thousands of queens every year. I think that is much more applicable to practical queen rearing. It is also a method that does not require grafting, good for those of us who can't see well enough to graft, and does not require the purchase of special equipment, good for those of us lacking in the funds to buy one of the graftless systems on the market.

I have also republished *Queen Rearing Simplified*, but from Mr. Smith's point of view this book, not that one, is the culmination of his work on rearing the best possible queens.

If you find any typographical errors or you wish to make comments please send them to bees@bushfarms.com

---

# Dedication

I dedicate this book to my wife, Ruth Reno Smith, who has faithfully stood by me through thick and thin and by her help and encouragement has made this book possible.

# Acknowledgements

It would be difficult to give credit to all who have helped me in the study of bees, but I must mention a few of those glorious saints who have gone on before: W.Z. Hutchinson, Dr. C.C. Miller, Geo. S. Demuth, G.M. Doolittle, Henry Alley, C.P. Dadant, Geo. W. York; dear old Prof. L.M. Kelchner, my art teacher of more than fifty years ago, who made the cover design of this book, and Dan Urbaugh, who induced me to go into commercial queen business.

Among the Living, I wish to thank my friend and neighbor George E. Judd, president of the firm of Judd & Derweiler, printers of National Geographic Magazine, for the excellent job of printing this book and to thank Mrs. Thomas Linger, the artist who made the cartoons.

A fine looking lot of cells, don't you think? Twelve in number. Count them. What! Only eleven you say? I will bet you forgot to count the dam-sel, Mrs. Huber F. Smith.

# Foreword

A philosopher has written, "Without regard to friends or foes, I sketch your world exactly as it goes." Now to sketch anybody's world and do an exact job at it is more than I wish to tackle. Nevertheless, I like the sentiment of the above lines so I shall adopt them in a revised form as follows: Without regard to friends or foes I shall attempt to sketch your world approximately as I think it goes. In other words, I shall go right ahead eating my grapefruit, let the juice squirt where it may. Moreover, I am writing this in my own crude way. Should I attempt to write in an approved rhetorical manner it would seriously cramp my style.

Moreover, I am writing this in the first person for I am telling just what I do with the bees. It is easier to say "I do so and so" than to say "It should be done so and so." I have greatly enjoyed reading the writings of Dr. C.C. Miller for he tells just what he did. In the introduction to his book *Forty Years Among the Bees* he writes, "Indeed I shall claim the privilege of putting in the pronoun of the first person as often as I please and if the printer runs out of big I's he can put in little i's." Incidentally, I wish to acknowledge the encouragement I received in my early beekeeping days from Dr. Miller who showed interest in my writings in the bee journals, as his card shows.

*Marengo, Ill. 3/6/19*

*My good friend,*
*The thing I don't like about you is your not doing more for the bee journals. You're one of the choice group whose writings I especially enjoy.*
*Blessings on you.*
*C. C. Miller*

I have always preferred to read authors who tell *how they do things* rather than those who tell *how things should be done*. Often there is a great difference.

# Table of Contents

# My First Recollection of Bees

My interest in bees dates back many years. I was born on a farm a mile southwest of Tampico, Illinois, October 13th, 1871, and one of my first recollections of that early life was watching my father in his shop making beehives. These hives were about a foot square and two feet high. The supers were mere boxes with glass on one side. When these supers were full they were taken off and honey, super and all were sold to the grocer.

I believe the shape of these hives was better than the shape of the hives in use today. In those hives the bees could better regulate the temperature, build up quickly and fill the supers. I believe my father made more net profit from his bees than most beekeepers make today, for those hives were intensely practical. He made the hives out of cheap material on rainy days. Of course nowadays, with foul brood prevalent, such hives could not be used.

When I was about three years old I remember king birds or bee birds, as they were called, used to catch my father's bees. Now he was a crack shot with the rifle, shotgun or revolver. I remember one day a bee bird was in the top of an evergreen tree catching bees. My father dropped the bird at the first shot using a .22 caliber Smith and Wesson revolver, the distance being about twenty-five yards. The bird's crop contained nineteen bees as I remember.

# Going West

In the spring of 1883 we moved to Dakota Territory. All my father's bees were sold at auction. We located near Sand Lake, some twenty-five miles northeast of Aberdeen. We took no bees with us but my uncle Perry Badgley took a number of colonies in box hives as he said he just could not get along without his bees. His bees all starved to death the coming summer as there was absolutely nothing for them to gather. Little did we dream that with the coming of sweet clover the Dakotas would be among the leading states in the production of honey.

## Getting the Bee Fever

While quite a small boy in Dakota I read a book on farm topics, and among them was an article on bees. As I remember the article it was not very authentic in its statements, but not knowing the difference I was fascinated with it and resolved that when I grew up I would have some bees. This idea never left me and after my marriage in 1899 and our removal to Vincennes, Indiana, I procured my first bees.

At that time I was running a private Business College, and as part payment for tuition, one of my pupils let me have two colonies of bees. They were in old-fashioned Langstroth hives with porticoes in front. That was in the spring of 1900. Eagerly I watched them all summer expecting them to produce a lot of honey. I had dreams of hot cakes and *my* honey. In the fall I cautiously peeked into the super after dark expecting to find at least enough fine comb honey to last us all winter. Alas, there was nary a drop!

Next spring there was but one colony alive and I began to realize there was something about bees I did not know, so with characteristic stubbornness I set about to learn something about them. I resolved that those bees should make me some honey or furnish a reasonable excuse.

## My First Swarm

This one colony built up rapidly and swarmed. It was Sunday and my wife was planning to go to church and had requested the pleasure of my company. I had shaved and was hoping something might happen to side-track the church-going when something did. Out came that swarm! Of course it had to be taken care of, so I got out of the church-going. The swarm clustered on a young peach tree and the tree was ruined.

Fortunately we had a little 3 ½ x 3 ½ Kodak and my wife took the picture shown.

*My first swarm*

## Beginning the Study of Bees

I then began in earnest to study bees. I got a hundred Danzenbaker hives. This hive had close fitting frames and violated everything Langstroth taught, or rather tried to teach. When those end bars got glued together they were solid as a rock. I used a wrecking bar as a hive tool. Years later that wrecking bar was stolen. I suspected a neighbor of stealing it. He had just bought some modern hives with long top bars! I put about a thousand dollars into those hives which later I found to be impractical. I burned most of them up. True, the manufacturer later apologized for inflicting such a monstrosity on the public, so we got apologies-but no reparations!

In spite of these financial losses, poor seasons and foul brood, I managed to increase the number of my colonies. As I grew in experience I learned to go more to the bee for my information. I learned that should I ask a dozen men a question I might get a dozen different answers, while if I asked the bees a question I got just one answer, and that the correct one.

No, they never gave me a five to four Supreme Court decision but their decision is always unanimous.

For Christmas 1902 my wife presented me with the bee book Langstroth on the Honey Bee, Revised by Dadant. This book set me on the right track and I still think it the best book ever published on bees.

## Go to the Bee Thou Sluggard

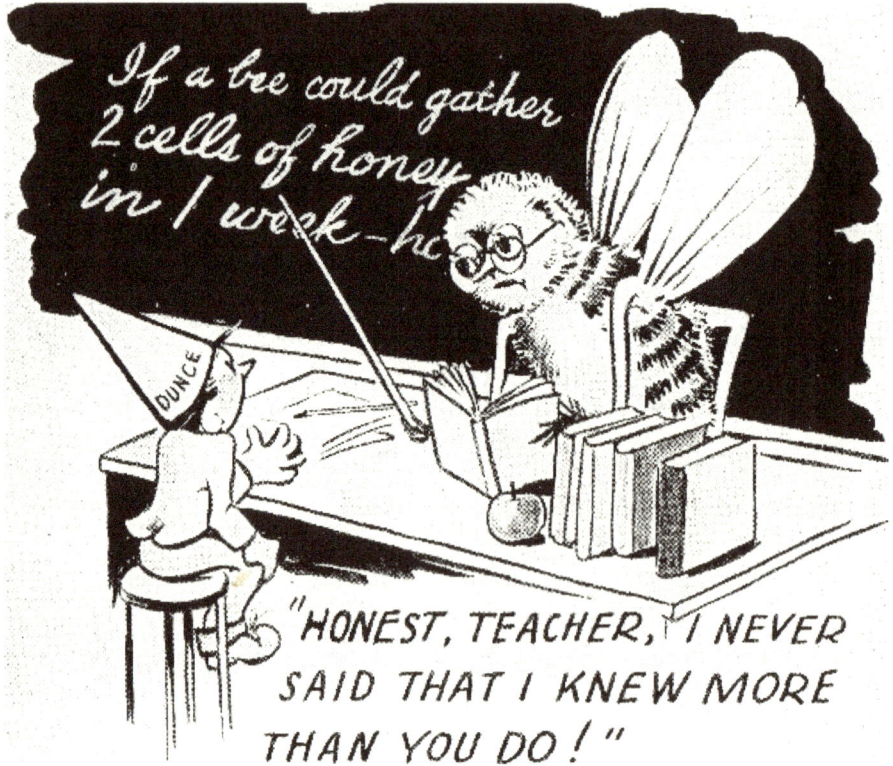

"HONEST, TEACHER, I NEVER SAID THAT I KNEW MORE THAN YOU DO!"

Solomon said, "Go to the ant thou sluggard. Consider her ways and be wise." I have often wished that Solomon had been a beekeeper instead of Samson being one for if he had been he never would have gone to the ant unless he carried a can of cyanogas. Nevertheless Solomon noticed something that many of us miss, and that is that insects have an instinct we might call an automatic intelligence which often is greater than anything in the head of man. The bees used the principle of evaporation for cooling the same as is used in our electric refrigerators and they used it probably long before man had learned to hang from a limb by his tail! However, Solomon rather squared himself with us beekeepers when he invited us to eat honey and acknowledged it to be good. So I tried to study bees themselves and learn their ways. I learned I could not force them to go my way but that I must go their way as far as is practical. Man is handicapped in that he must go to school and study in order to learn anything, while the bee is born with its college diploma in its hand. The bee's education is complete. They agree on all points while if you can find two men who agree on anything you are to be congratulated. The bees know all they need to know and you cannot teach them any more. I learned I must be the pupil and the bee the teacher.

I made a careful study of queen rearing and bought breeding queens from Doolittle and Alley and also imported some Italian queens. One queen I received from Mr. Doolittle was a wonder. Her bees were of a beautiful yellow color, very gentle and splendid workers. So pleased with her was I that I used her for the foundation of our stock. This decision I have never regretted.

## Beginning Commercial Queen Rearing

One day in 1912 State bee inspector, Dan Urbaugh, came to inspect my bees. I remember he stood for some time in silence watching the flight of the bees. Then he turned to me and said, "Why don't you sell some queens?" Then he said, "Those are the finest bees I ever saw." I asked him how he could judge bees by merely looking at them. He replied, "I can tell by the looks of them and by the way they act. Here I am right among them and none offers to sting." Having had little experience with bees except my own I took it for granted all bees were like that. I told Dan that I would greatly enjoy rearing queens if I could sell them. I shall never forget how he looked at me and replied, "Sell them, why of course you can sell queens like these and you will not have to do much advertising either." I rather had my doubts but thought I would try it. I made some mating hives of various models and when I had a dozen queens on hand I wrote Mr. Urbaugh I had some queens ready. I made the remark to my wife that maybe Mr. Urbaugh was just talking to please me and really did not mean what he said. In a few days I received a check from him for six queens. I said to my wife, "I thought he was just kidding, but that check doesn't lie."

Thus at Mr. Urbaugh's request I became a full-fledged commercial queen breeder. I could have gone into some other business and no doubt made more money, but doubt if I would have gotten so much out of life. I have given all there is in me to the thought of better bees and especially better methods of producing them. During the years that have passed since 1900 I have spent many thousands of dollars in experimenting. When I found a new mating hive that suited me better than the ones I was using I would burn the old ones in the furnace. As Josh Billings put it, "Experience is a gude teacher but the tooishun kums purty hi."

## Short Course in California

One of the outstanding experiences which helped and encouraged me in my work was during World War I. I worked as extension agent under Dr. E.F. Phillips who was then in charge of the bee work at Washington. Dr. Phillips, Geo. S. Demuth, Frank Pellett, M.H. Mendleson, Dr. A.P. Sturdevant and myself conducted a short course in beekeeping covering most of the state of California. The flu was rampant and we wore flu masks. We also were requested to sneeze in our handkerchiefs in order not to spread the germs. However, I learned it took a professional bad man to be quick enough on the draw to get the drop on a sneeze after he felt one coming on! I have often said I do not know whether or not I taught the California beekeepers anything worthwhile, but they taught me a great deal. One thing stands out in my mind most vividly and that is what a splendid lot of men and women it is who keep bees. I doubt very much if any other industry can show a better and more honorable set of people.

# Queen Rearing Simplified

In 1923 my very good friend Geo. S. Demuth, asked me to write a book on queen rearing and *Queen Rearing Simplified* was the result. It was the best that I knew at the time but with the passing of more than twenty-five years we have made so many radical changes for the better when quality queens are desired that now we use practically nothing described in its pages. However, it has given me much satisfaction to receive letters from beekeepers all over the world thanking me for the help given in that book. It is my belief that *Better Queens* will receive a greater degree of appreciation. If this is true I shall feel amply paid for all the effort and expense I have been to in preparing this volume. With the passing of the years there may be minor changes in our system of rearing queens direct from the egg but the main feature will always remain if quality of queens is desired. With our present system all cells are as large and as well supplied with bee milk as are the cells produced by the bees during swarming or supersedure in nature, and let no one tell you he can beat nature in rearing queens. All we claim is that we can equal nature and that is enough.

(Link to Queen Rearing Simplified:
http://www.bushfarms.com/beesqueenrearingsimplified.htm )

# Shall the Beekeeper Rear His Own Queens?

My answer to that is emphatically *yes*, providing he can rear *the best of queens*. Maybe some member of the family, a boy or girl or yes, maybe Mother, may rear the queens and find it not only the most profitable work about the beeyard but they will get a world of pleasure in doing it. Many will make a larger profit from half the number of colonies if they rear their own queens. Therefore, instead of being more work to rear their own queens it really would be less work for the time saved in caring for double the number of colonies not to mention the larger amount of money invested. Let me give just one instance. In California I had the privilege of observing the systems of two beekeepers who worked in opposite lines. One had two thousand colonies. When a colony died from becoming queenless the beekeeper divided another colony. In one division was the queen. In the other, the bees were expected to rear a queen. Sometimes they did, but having only old combs the resultant queens were not the best. He had to hire a large crew of men to take care of all these colonies to see that they did not die out. In addition, the interest on the money invested in that large equipment was considerable and if you consider the depreciation, that was an additional expense.

In contrast to this I met a man in Ventura County who had but 250 colonies. This man and his son reared their own queens and requeened every year. He told me that year their colonies averaged 200 lbs. per colony and they sold the honey wholesale for 22¢ per pound. It was plain this man and his son made more clear money from the 250 colonies than the man who operated the two thousand.

## How Bees Rear Queens in Nature

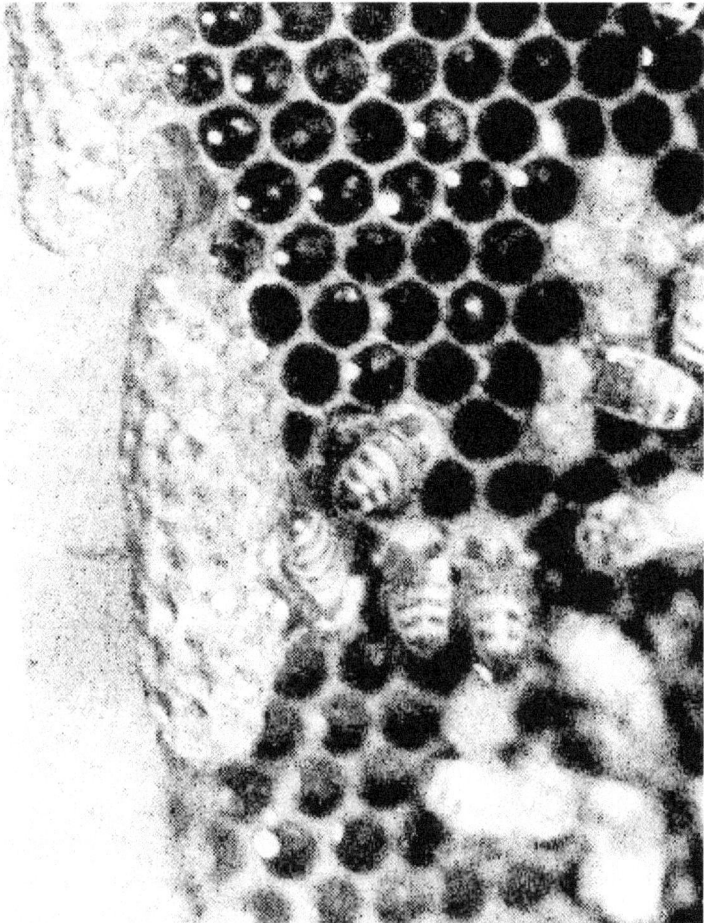

*Swarm cells*

As stated, let us now lay aside all man-made theories and go straight to the bees and see how they do it, for we must remember that if we are to rear the best of queens we must follow the bees in their work and duplicate their work insofar as it is practical. We must remember the proven fact that Mother Nature knows how to rear her own while man may use cheaper methods he never can improve on nature when *quality* of queens is considered.

When bees rear queens they do it in one of three ways-first, when preparing to swarm; second when they wish to supersede a failing queen; and third, when a colony becomes queenless for some reason such as an accident to the queen. In such a case they rear a queen by what we may term the "emergency" method.

## Queens Reared During Swarming

When there is an abundance of nectar and pollen coming in from the fields and the colony is built up to good strength with plenty of brood in all stages in the hive and the bees become crowded, they may decide to swarm especially if the queen be an old one. As all colonies do not act alike some may swarm while others will not. The first step in preparing to swarm is the building of queen cells. The bees prefer to build the cells at the bottom of the combs or at the sides if there is room. These cells are pointed downward.

After the cells are nicely started the queen lays an egg in each. In about three days the eggs hatch and the larvae are surrounded with an abundance of bee milk. Sometimes the milk is placed in the cell before the egg hatches. The larva grows at an amazing rate. It has been stated that if a human baby grew as fast as a queen larva and weighed ten pounds at birth, in five and one half days it would *weigh twelve and a half tons!*

By the time the cells are sealed they contain nearly half an inch of dried bee milk. Now please consider the fact that if we wish to rear as good queens as the bees do when preparing to swarm we must duplicate their performance as nearly as possible.

In nature, when swarming, the bees seldom need more than one cell and never more than three or four, but they often build from one dozen to two dozen. Why this extravagance in cell-building I do not know but possibly the bees are secreting so much milk in their glands they want to get rid of it. This desire to get rid of this over-supply of milk is probably one of the causes of swarming. I have often prevented swarming by adding unsealed brood from other colonies. This gave the bees so much larvae to feed that they found an outlet for the excess of bee milk and the colony tore down the queen cells it had started.

## Queens Reared During Supersedure

In case the queen is getting old or, as is all too often the case when the grafting method is used, and the larva is underdeveloped, the bees may ask the queen to retire as they want to put a young vigorous queen in her place. In doing this the procedure is the same as in swarming but not so many cells are built. Usually from three to five. Conditions and locality may cause a variation in this matter.

The first virgin out will tear down the remaining cells unless the colony decides to put out an after-swarm. In due time the virgin flies out and mates with a drone and in two days she begins to lay. Sometimes either the bees or the new queen kill the old queen. At other times it appears that the old queen is sensitive or proud and thinks she is not properly appreciated and she with a few of her close friends simply up and pull out. This is called a supersedure swarm. Sometimes this swarm doubles up with another. I have known this to happen and the old queen was accepted and both queens carried on for some time. These old queens are readily accepted by the bees. I have dropped an old queen into a colony of laying workers where she was accepted and in time she reformed the laying workers and later the queen was superseded. The length of the life of a queen depends largely on the number of eggs she lays. In the North where the queen has a long rest period in winter she will live longer than she would here in Florida where she lays the year round. We find that for best results here in Florida the colony must be requeened every year. In the North if the queen is doing good work the second year she might be allowed to live the third year but as a rule two years is long enough to keep any queen. Any queen not doing good work should be removed regardless of her age.

## Queen Alice

In Indiana we had a queen we named Alice which lived to the ripe old age of eight years and two months and did excellent work in her seventh year. There can be no doubt about the authenticity of this statement. We sold her to John Chapel of Oakland City, Indiana, and she was the only queen in his yard with wings clipped. This, however is a rare exception. At the time I was experimenting with artificial combs with wooden cells in which the queen laid.

## Queens Reared by the Emergency Method

In case the bees are left alone and not interfered with by man, probably it is very seldom that this method is ever used. Where man interferes and opens the hive frequently the queen may be killed. In such a case, provided there are larvae of the proper age, the bees build queen cells over some of the larvae in the worker cells, such larvae originally intended for workers. It has often been observed that many of the queens reared by this method are not the best. It has been stated by a number of beekeepers who should know better (including myself) that the bees are in such a hurry to rear a queen that they choose larvae too old for best results. Later observation has shown the fallacy of this statement and has convinced me *that bees do the very best that can be done under existing circumstances.*

The inferior queens caused by using the emergency method is because the bees cannot tear down the tough cells in the old combs lined with cocoons. The result is that the bees fill the worker cells with bee milk floating the larvae out the opening of the cells, then they build a little queen cell pointing downward. The larvae cannot eat the bee milk back in the bottom of the cells with the result that they are not well fed. However, if the colony is strong in bees, are well fed and have new combs, they can rear the best of queens. And please note-- they will never make such a blunder as choosing larvae too old.

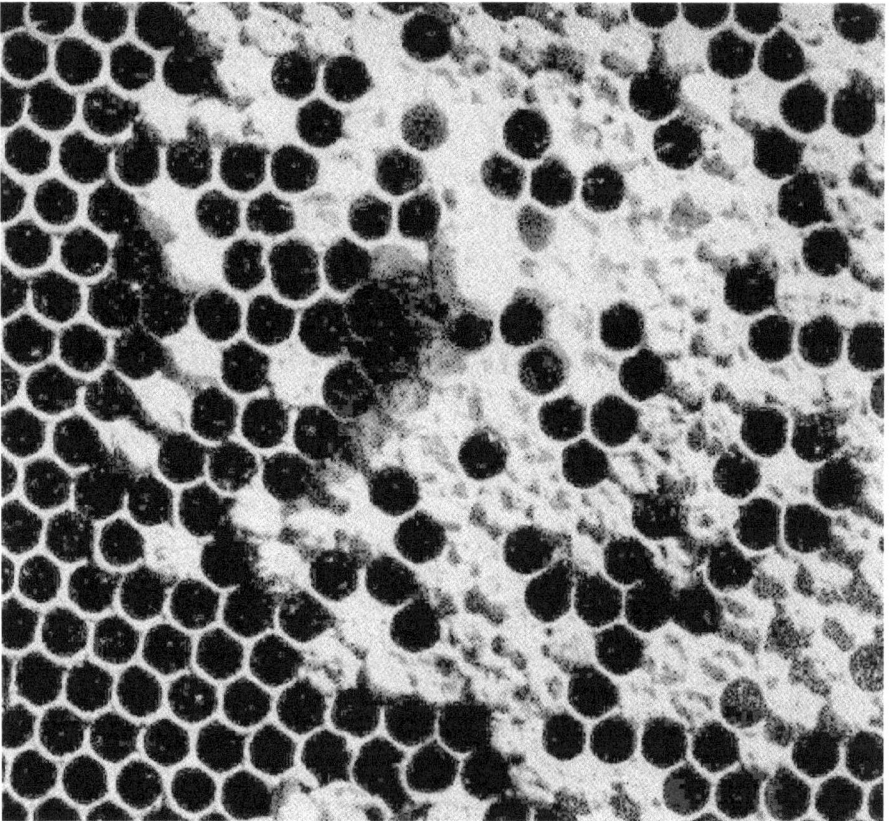

*An emergency cell*

## Dr. Miller's Method

In reviewing the past it is interesting to note how many really good beekeepers discarded the grafting method after giving it a trial. A notable example is Dr. Miller. His method will produce just as good queens as can be produced. He gave a piece of new comb to the colony containing his breeding queen. As soon as the eggs hatched he trimmed the comb back to the larvae and gave it to a strong colony made queenless and broodless. The bees could easily tear down the new combs and build queen cells. For the person wishing to rear just a few queens his system is good. It would not do for the commercial queen breeder as too many cells are built together and it is difficult to cut out the cells without injuring them.

Another disadvantage to his system for cell building on a large scale is that the bees do not start as many cells with his method as they do with our system. After all it is little more work to mount the cells on bars as we do and we get equally good cells and more of them. Now please remember the *quantity* of bee milk the growing larvae receives is just as important as *quality* for this will be mentioned frequently in the following pages.

(Link to the Miller method:
http://www.bushfarms.com/beesmillermethod.htm )

## Artificial Queen Rearing-Its Development

It is not clear who first practiced grafting, or transferring larvae from certain cells to other cells. As far back as 1791, Francis Huber, the naturalist wrote: "In a hive deprived of its queen I caused to be placed some pieces of comb containing eggs and already hatched worms of the same kind. The same day the bees enlarged some of the worm cells and changed them into royal cells giving to them a thick bed of jelly. We then removed five of the worms placed in these cells and Burnens substituted for them five worker worms which we had witnessed hatching from the eggs forty-eight hours previously. Our bees did not appear to be aware of the exchange." (From *Huber's Observation on Bees* translated from the French by C.P. Dadant.) Therefore, the grafting method is more than 150 years old. Since that day many have placed larvae from superior queens into the cells of inferior queens.

(Link to Huber's book:
http://www.bushfarms.com/huber.htm )

## G.M. Doolittle

The greatest impetus to the grafting method was given by G.M. Doolittle sometime in the 80's when he devised the method of dipping queen cells. Before that time the queen breeder had to depend on the few cells he could find in his hives. Some took soft wax and molded it around a stick making a crude cell, and these were offered for sale at a penny apiece. Mr. Doolittle, in his excellent book *Scientific queen-rearing*, states it came to him, "Why not dip the cells the same as my mother used to dip candles?" Soon after that Henry Alley brought out his system of rearing queens direct from the egg, the system on which ours is founded.

(Link to G.M. Doolittle's book:
http://www.bushfarms.com/beesdoolittle.htm )

## The Grafting Method

The object of *Better Queens* is to be helpful to all who rear queens and not to criticize those who use the grafting system. After all, I am criticizing the method I taught in *Queen Rearing Simplified*, so it is perfectly legitimate to criticize oneself! Many who now are using the grafting system and who want to rear better queens will want the two systems compared. As most bee-keepers know, by the grafting method we mean the method in which the larva from a worker cell is transferred to an artificial queen cell. We used that system for 33 years. Not one of those years did we get the fine large cells which are necessary to produce full developed queens throughout the whole season. We found that when there was a light honey flow with plenty of pollen coming in, and if we kept the cell builders up to great strength, we could get a very high percentage of good queens. Even at its best we had to cull cells and virgins and frequently to discard laying queens that were not fully developed. Even then a few inferior queens would get by us which we had to replace. This never happens with our present system. We never have thrown away a cell for being too small, for all are alike. With the present system we have yet to see an undersized virgin. When using the grafting system, when there was no flow, it was well-nigh impossible to get good cells even though we fed sugar by the ton. Not one of those 33 years passed in which I did not long for a system with which I could produce those fine large cells which I had observed in colonies preparing to swarm, a system by which I could produce cells in quantities throughout the entire season.

## The Desire for Better Cells Intensified

The wish for better cells was greatly enhanced while we were having a beekeeper's picnic at our place in Vincennes. I was asked to demonstrate queen rearing while using the grafting method and naturally wanted to make as good a showing as possible. As was our custom, I had built up several colonies to great strength by adding frames of brood and bees from other colonies. One colony of great strength swarmed just as the picnickers arrived. Our State Inspector, Yost, weighed the swarm and reported it weighed exactly 25 pounds. It looked as though as many more bees remained in the hive so they estimated that colony must contain at least *two hundred and fifty thousand bees* - many more of course, than could be the product of one queen. I was proud of the cells the colony produced, and for grafted cells they were fine ones. It so happened that this colony had built some swarming cells in the brood nest below the excluder. It has been stated that "Pride cometh before the fall" and I will say that my pride took an awful rumble when I saw those enormous swarming cells. I realized I did not know the first principle about raising really good cells. *Their cells were nearly twice as long as mine and had at least three times as much dried milk in them as did mine.* I realized as never before the shortcomings of the grafting system if *quality* of queens is desired. However with the inborn stubbornness for which I am noted, I resolved never to quit till I had a system in which I could duplicate the performance of the bees in building cells while swarming.

## Fifteen Years of More Work

It took fifteen years of study and experimenting before I got the system perfected with which I can duplicate the work of bees building cells while swarming. At last we have it and we have made very few changes in it for several years, and I doubt if we ever will make any important change. At present, most queen breeders are using the system I taught in *Queen Rearing Simplified*. I feel confident most of them will adopt the new system we are teaching in *Better Queens*, for buyers of queens and package bees are beginning to demand better queens - queens that will not be superseded almost as soon as they begin to lay, thereby losing a crop of honey.

## What the *American Bee Journal* Has to Say About Poor Queens

In October 1947 number of the *American Bee Journal* appeared this article:

"This is the season of disastrous queen failures. From every direction come reports of supersedure and queenlessness on an unprecedented scale. With an abundance of clover the bees have failed at a very critical time. The failure of the queens results in heavy loss to the colony. When egg laying ceases for ten days to two weeks in the spring much of the productive force is lost. Losses have been especially heavy with package bees. The cold and backward spring was unfavorable for replacement and in too many cases the colony has been lost entirely. Most certainly we need to know more about the reason for this serious condition which has cost the bee-keepers a substantial portion of the possible 1947 Harvest."

## Yes, We Need to Know More

So let me tell you. It is safe to say that 95% of the cases of premature supersedure are caused by *improper feeding of the young queen larva*. The grafting system simply will not produce the best of queens. A package shipper who uses queens reared direct from the egg reported to me that he did not have a single complaint about premature supersedure but, on the contrary, had scores of letters praising the performance of the queens and the splendid work the package bees did. Queens must be reared direct from the egg and the finishing colony must be fed *honey* for best results.

## Experiments

Let us look at a few of the experiments I have carried on, many of which now look foolish. Mr. Eugene Pratt wrote a pamphlet entitled "Forcing the Breeding Queen to Lay in Artificial Queen Cells." Had I understood bee nature then as I do now I would never have tried it, for when he said "forcing" that would end it right there. I got the outfit and tried it. The wooden cups were corded up into a supposed comb and the queen was supposed to lay in it, and she did - about a dozen eggs then she went on a slow-down strike, then a sit-down strike, and finally a walk-out. The few cells I did get were very good but still were not flooded with the bee milk as I wished. So once again I sorrowfully returned to the grafting method.

Next I gave the bees only drone comb, as it is known the queen will lay worker eggs in drone cells if there is no worker comb present. The results were not quite so bad, still I could not get the bee milk into those cells in the quantities I knew were necessary to produce the best of queens. Then I had some wooden cells made smaller in size and square on the outside so they would cord up more like cells in a worker comb. It looked much like a comb of worker cells to me. Evidently it did not look that way to the bees or the queen for, after laying a few eggs, she went A. W. O. L. and had to be repatriated. The cells and the resultant queens were nothing to be proud of. I had not yet learned that *quantity* of bee milk was just as important as *quality*, and I just could not get the quantity of the bee milk into those cells. The reason was that it takes a large number of bees to provide the milk in abundance and it was not practical to use a large number in such experiments. Once more I reverted to type and went back to grafting.

## The Alley System

Some may ask why I did not use the system used by Henry Alley in which the queen laid in the worker cells. I had tried his system and failed with it as many others have done. Mr. Alley recommended that we use combs in which bees had been reared as they were so much easier to handle, especially in hot weather. True, the combs are easier for *us* to handle but not so easy for the bees to handle, and as the bees have to do the work they should be consulted in the matter. I tried Mr. Alley's system using old combs. The bees did not accept the cells as readily as I wished and the resultant queens were no better than the ones reared from grafted cells. I learned later that the reason for the failure was that the bees could not tear down the cells and rebuild them into queen cells on account of the cocoons in the cells. They had to fill the worker cells with bee milk, floating the larvae out to the mouths of the cells much the same as when cells are built under the emergency system.

(Link to Alley's method: http://www.bushfarms.com/beesalleymethod.htm )

## Cells Built From New Combs

In the year 1934 I observed a *strong colony* in a hive containing only *new white combs*. They were hybrid bees and I had removed their queen intending to introduce a better queen. I had to ship out so many queens that I put off requeening this colony. We finally had a queen to introduce and on examining this colony I was astonished to see large well-developed queen cells such as I had seldom seen. Upon further examining this colony I found that they had torn down the comb and a number of worker cells and had built these fine cells over the very small larvae in the cells.

Then the thought came to me, why not try the Alley system using all new white combs? I went to my best breeding queen, removed all brood and placed a new white comb between two combs containing honey and pollen. The queen immediately began laying in it. After 24 hours this comb containing eggs was placed over a strong colony above a queen excluder. As soon as the eggs hatched I cut the comb in strips, mounted the strips on cell bars so when given to the bees the cells would point downward. Huber then removed the larvae in two cells leaving a larva in one till two bars were prepared.

## On the Road to Success

These prepared cells were then given to well-fed bees confined in a starter hive. This was Saturday and I lay awake late in the night wondering if it would work. In fact I lay awake all through church the following Sunday pondering over this question. Would the bees accept the cells? If so, would they build them out sufficiently so the bees that were to finish them would recognize them as queen cells? Maybe they would tear them down or just finish them as workers. I had intended leaving the cells with the bees in the starter hive for the entire 24 hours but curiosity got the best of me, so right after church I looked at them.

# Hurrah! Success at Last

When I examined those cells it was by far the most inspiring sight in my beekeeping life. Not only was every cell accepted and well developed but the larvae were fairly flooded with bee milk although they were so small as to be visible to only one having excellent eyes. They were much too small, even at this age, to be used for grafting. I cornered each member of the family and exclaimed (my wife says I yelled) "Here is where we discontinue grafting." No one can realize the joy and satisfaction that was mine as I realized at last I had reached the goal I had vainly tried to reach for more than a quarter of a century!

I realized there was yet much to be learned before I made the system practical so I could produce cells in abundance all through the season. However, I knew this could be done for the main object had been attained. I have tried many plans over a period of ten years or so now I see no chance for improvement.

# The Difference Between a Queen and a Worker

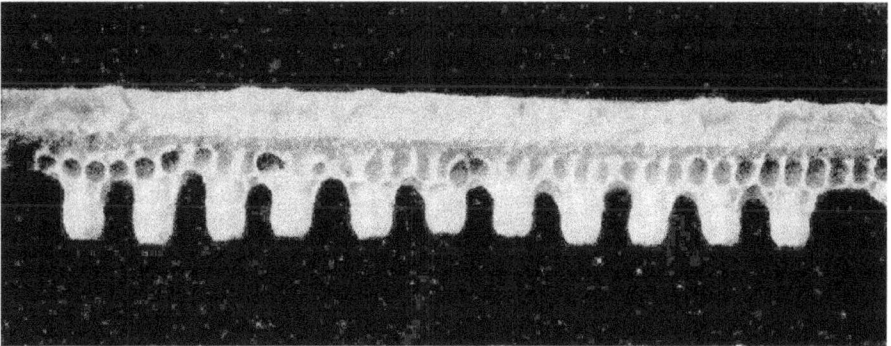

*My first direct-from-the-egg cells*

The difference between a worker and a queen is brought about by the different food each receives and the size of the cell it is raised in. Both are hatched from the same kind of egg, that is, a fertile egg. The drone is hatched from a nonfertile egg. Right here is an interesting phenomenon. Just how can the queen lay fertile eggs in the worker cells and nonfertile eggs in the drone cells? This she does with great regularity. A queen that has not mated with a drone will lay eggs but they will produce only drones. But to return to the question as to what makes a queen instead of a worker. Scientists who have done considerable work along this line tell us that the food of the worker larva and the queen larva is the same for approximately 48 hours. This food is bee milk secreted in the glands of the worker bee, such glands being located in the head of the worker. The discovery of these glands is attributed to Meckel, who discovered them in 1846. After about 48 hours the food of the worker larva is changed and it is fed honey and pollen. The queen larva is fed bee milk all through her larva life which is about five and one half days. Many have missed the all-important point and that is the queen larva receives *more of it.* And if we wish to produce the best of queens we must bear in mind that the *amount* of food the queen larva receives is fully as important as the kind of food it receives.

## One of the Great Miracles of Nature

Let us see what a miraculous food this bee milk really is. The queen larva is given this food three and one half days longer than is the worker larva. This three and one half days' diet has made an entirely different bee out of it. Some have stated that there is nothing really miraculous about this. The queen being fed on more nourishing diet, develops into a fully developed female while the worker larva, being fed on less nourishing diet, is dwarfed in its growth. If this were true it would simplify the matter, but it is exceedingly far from the truth, for the worker bee is developed much more fully in a number of points. I believe I am the first to assert that the worker bee is as much a perfectly developed female as the queen but is *developed along different lines.*

## The Worker Bee as Fully Developed a Female as the Queen

Let us look into this matter and note a number of features in which the worker is more fully developed than the queen. In the first place the worker has milk glands in its head in order that it may nurse the young larvae *and these glands are lacking in the queen.* If you ask most any mother I believe she will tell you that it is as much of a mother's job to nurse the baby as it is to bring it into the world. Yes, sir, when it comes to rearing bee babies the worker bees have the heavy end of it. It would be just as logical to say that the three and one half days' diet the queen received has dwarfed her so she cannot nurse her own babies.

No, neither worker nor queen is dwarfed, but they are entirely different bees both in physical make-up and temperament. The worker is more fully developed than the queen in a number of features. The worker has pollen baskets which are lacking in the queen. The worker has barbs on its sting while the queen has none. The sting of the queen is curved while that of the worker is straight. In all of this one can see the wonderful work of the Creator. Probably the reason for the queen's sting being curved is that when she is full of eggs it would be difficult to bring her battery to bear on her rival if her sting were straight. The worker bee has wax glands and a honey stomach, both lacking in the queen. The head of the worker bee is larger than the head of the queen, according to Snodgrass. There are several other slight differences between the worker and queen, all brought about by a change in diet for three and one half days.

Again the nature of the queen and the worker is entirely different. The queen will never sting a human being, while if you think the workers will not, you come with me. As stated, a queen will never sting anything but a rival queen. I might qualify that statement by saying a queen never stings anything but a queen, *or what she thinks is a queen.* I was stung by a queen once but I insist it was a case of mistaken identity, for she thought I was a queen. It happened thus: I had been requeening some colonies and in removing the old queens I killed them by pinching them between my thumb and finger. I had wiped my thumb and finger on my trouser leg. A virgin queen circled me a few times probably to adjust her bomb sights then made a pin-point landing on the spot where I had wiped my thumb and finger, and planted her sting in my leg. Yes, she thought I was a queen. While greatly appreciating the compliment, I would much prefer she would show her appreciation in a less militant manner.

Now as the queen and worker larva receive the same kind of food for the first two days, the reasoning has been that if larvae two days old are placed in artificial queen cells, perfect queens can be reared. As the attorney would say, "Your honor, I object." "While the larvae may be fed the same kind of food for the first two days they have not been fed the *amount* of food they would have been fed had they been reared in their own cells direct from the egg." "Objection sustained."

## For Best Results Follow Nature

We should note that when bees are preparing to swarm or supersede their queen they fairly flood the young larvae just as soon as the eggs hatch and sometimes they put bee milk into the queen cells before the eggs hatch. This they do in such a lavish manner that after the bee has emerged there remains from a quarter to half an inch of dried milk. (I suppose we now should say "dehydrated milk.") This is the way Nature rears her queens and don't say Nature makes a mistake by putting in too much food. If this overabundance of milk were not necessary it would not be put there.

Therefore, if one does not have his cells built so there is a great abundance of dried milk left in the cells after the virgins have emerged, *he is not rearing the best of queens.*

## Shortcomings of the Grafting Method

When Dr. Phillips was head of the bee culture laboratory in Washington, he wrote me that the word "grafting" was an improper term and that they did not graft but transferred. I wrote him it was refreshing to know there was one department in Washington that did not graft! In using the grafting method the larvae are left in the worker cells for two days where they are sparingly fed, for the bees are making workers of them. If much younger larvae are used they will perish, for they cannot stand such rough treatment. If you will examine the larvae two days old you will see very little bee milk around them. In fact, they are being "rationed." My experience has proved without a shadow of a doubt that such larvae have been starved in such a manner that they will never become fully developed queens no matter how lavishly they are fed after that. But wait-that is not the half of it. In grafting, you take the larva away from the starvation ration it has been getting and place it into an artificial queen cell, unless you have committed hara-kiri on it in the operation. Things are already bad enough, but putting it into that artificial queen cell!!-well, that is just about the blade of timothy that fractured the spine of the dromedary! We used to prime our cells with bee milk but, after careful examination, believe it was a detriment, for the first thing the bees do is to remove all the milk we had put in. Grafting in bare cells is better-or rather not so bad.

# What Promised to be a Bright Idea Proved to be Very Dim

In order to get the cells filled with bee milk the same as they are when built during swarming, I allowed the larvae in the grafted cells to remain for two days till there was plenty of bee milk in them, then removed the larvae and put in young larvae. I hoped to get fine cells in this manner but the bees seemed to think otherwise. They accepted but a few of the cells and in some cases the larva was pushed over to one side of the cell and the bee milk all removed. In a few cases the bees accepted the cells but placed a little thin milk on top of the milk already in the cells. The few queens reared were no better than the ones reared by grafting in the regular way. This leads me to believe the dried bee milk is not suitable for larva food but rather is the crumbs left over from the feast. I believe the real food is the very thin milk that is fed to the larva. Then one may ask what is the advantage of having so much dried milk left in the cells. That merely indicates that the growing larva was fed in a lavish manner which is very necessary if quality of queens is desired. One queen reared as described above performed in a manner I never knew one to perform before or since. She laid drone eggs only but none of them died in the cells as is common with drone layers but all matured into perfect drones and that queen fairly filled that hive with beautiful drones. Scientists who have made a study of the subject tell us that bee milk is the same in all stages. I am inclined to doubt this. It may be changed in the moisture content only but I have observed that bee milk in the cell of a larva of one age will not be accepted by a larva of another age. When given, the bees immediately remove it and proceed to give the proper nourishment.

Many of our best beekeepers abandoned the grafting method after giving it a trial. I once wrote a paper, which was read at a meeting of the National Beekeepers Association, setting forth the shortcomings of the grafting method, although I was using it as the time. Dr. Miller wrote, "I say Amen to that."

# Bees in All Stages of Development from Worker to Queen

Many seem to think there is a marked division between worker and queen. They seem to believe that the bee is either a perfect worker or a perfect queen. I believe I am the first to point out that bees can be reared in all stages from a worker to a queen and these stages are brought about by the *amount* of bee milk the growing larva receives. These different stages are not only brought about by the kind of food the larva receives but, don't forget, by the amount it receives. In our work we have observed bees in many stages as stated above. Upon a number of cases we have observed a bee like a worker but with a more pointed abdomen with a yellow color like that of a virgin. Such a bee remains in the colony and acts like a worker and in no way interferes with the queen cell that is given or with the virgin when it emerges. Whether or not such a bee goes to the fields or performs any useful work about the hive I have never been able to determine. Then there is a bee a few steps above that that is the meanest most contemptible and the most (censored) bee that ever existed. She is similar to the bee just described but a little larger and more like a queen but so small one might not notice her even tho she be in plain sight. She will tear down cells just as fast as you give them. What makes her such a pest is the fact that she occurs so seldom that you are not looking for her but are trying to find a regular virgin.

I do not know where such a bee ever flies out to mate, for when I finally get sight of her she never gets a chance.

A bee a few steps above the one described flies out to mate but never gets back. The bee above the last-mentioned flies out and mates, comes back, lays a few eggs and is superseded. All too many of such queens are sent out by breeders using the grafting method. A queen above the last may remain in the hive and lay sparingly, thereby losing a crop of honey and either dies in winter or is superseded early next spring. Then there are queens from this on up to the fully-developed queen that keeps the hive packed with worker bees which produce a bumper crop of honey. The first bee mentioned is reared in a worker cell while the next may be reared in a small queen cell that may often be taken for a drone cell.

## The Mating Flight is Not for Weaklings

The mating flight of the queen is a very strenuous affair and it is quite evident that Nature made it that way in order to eliminate the unfit. When a dozen or a thousand drones take after the queen she flies in a zigzag manner so that only the strongest flying drones can overtake her. Upon one occasion when the queen and a horde of drones were in the air they made such a buzzing that Mrs. Smith asked if that was a swarm. It sounded like one. In mating, quite frequently both the queen and the drone fall to the ground. After the exhausting chase the queen will be unable to take wing again unless she be possessed with great vigor. While using the grafting method in rearing our queens we took it as a matter of course that a certain percent would be lost in mating. This loss we now know was caused by rearing weak queens. With our present system of rearing queens direct from the egg a virgin queen in the hive invariably results in a laying queen. In mating, the drone soon dies. It affects him much as it does the worker in losing its sting. The cut shows a worker bee removing the drone organs from the queen. This is a very rare photograph and was taken purely by accident.

*Worker bee removing drone organs from the queen.*

# Sombre Side of the Romance of the Honey Bee

The honeymoon of the bee is short for no sooner does the queen become a bride than she is a widow, and no sooner does the drone become a groom than he passes away. There is one advantageous feature in this in that the bride is saved a trip to Reno. Possibly when the groom realized he is about to become the father of half a million children, and every blasted one of them females, it is more than he can endure. A queen that has never been to the altar will lay eggs but they will produce drones only. Hence the drone has no father but does have a grandfather on his mother's side. Neither does he have any sons but may have many thousands of grandsons. However, they never get to greet Grandpa for he has been gone months before they were born.

# Some Complications in the Bees' Domestic Affairs

When a yellow queen mates with a black drone all her sons will be yellow like herself while her daughters will be partly black. Now the question arises, are these pure Italian drones and half-breed German bees full brothers and sisters? Can a drone and worker be full brother and sister? I wonder.

# Food and Yet More Food Makes the Perfect Queen

The different grades of queens we have mentioned, from the poorest to the best, are brought about by the kind of food the larvae receive and what is of equal importance the *amount* of such food. If we wish to produce as good queens as the bees produce during swarming and supersedure, and don't let anyone try to make you believe he can do better, we must duplicate the bees' performance as nearly as possible. When building a swarming cell, after it is well started, the queen lays an egg in it. Just as soon as the egg hatches the bees fairly flood the cell with bee milk and they keep up this abundant feeding until the cell is capped over. From this it can easily be seen that the grafting method is entirely unsuited for rearing queens if quality queens are desired. Four or five pounds should be kept in the breeding hive. As this large force of bees have no brood to feed except the small patch in the new comb given *such larva receive exactly as good feeding as do the larvae in swarming or supersedure cells.* From what has been said it is evident that in order to produce fully developed queens that will keep a large brood-nest packed with bees the larvae must have an abundance of bee milk just as soon as they come from the eggs and *must have an abundant and constant feeding without interruption until the cells are sealed.*

## Importance of Food for the Young

Let us look at the matter of food for the young. The newly born of any specie must have an abundance of food if proper development is obtained. Mature creatures may go on a fast for a period without injury and in many cases such a fast may be beneficial. Not so with the young. Many people grow old prematurely because they were given artificial food when infants instead of their mother's food. Many babies die because some quack doctor wants to improve on Nature and prescribes some new-fangled food rich in vitamins from A to Z. I know from sad experience. I believe if the truth were known there are many now behind prison bars who would not be there if they could have had their mother's milk for the first six months of their lives. For lack of proper nourishment their bodies and brains were not properly developed. I have observed the effects of food on the young of animals and fowls for many years and the more I observe these things the more I believe in proper and abundant feeding of the young.

## Honey as Food for Man and Bees

It has been proved that the worst foods for man are white flour and sugar. Therefore, if sugar is unfit for human food, how much more unfit is it for food for bees? Honey is the food God made for bees as bees produce more than they need it is evidently intended that we should have the surplus. Honey is rich in minerals and valuable enzymes which are totally lacking in sugar.

I have proved to my satisfaction *that the best and most vigorous queens cannot be reared when sugar is fed to the cell-building colony.* Only honey should be fed if you wish to rear those prolific long-lived queens. To prevent robbing when feeding honey a feeder should be placed on the back of the cell-building colony the same as that on the breeder hive.

It is well known that improper diet makes one susceptible to disease. Now is it not reasonable to believe that extensive feeding of sugar to bees makes them more susceptible to American Foul Brood and other bee disease? It is known that American Foul Brood is more prevalent in the north than in the south. Why? Is it not because more sugar is fed to bees in the north while here in the south the bees can gather nectar most of the year which makes feeding sugar syrup unnecessary?

We feed sugar syrup to the bees in the mating hives in order to get them to accept the queen cells but as the virgins are yet in the cells no harm is done. If the truth could be learned I believe it would be found that as soon as the virgin emerges from the cell it would go to the cells containing honey rather than to those containing sugar. The reason we do not feed honey in order to get cells accepted is that it would cause robbing, especially in the mating hives where the bees are few in numbers. Even at that we feed only when no nectar is coming in from the fields. Of course if disease is present honey should not be fed unless it is heated to destroy the disease germs. Heating would destroy the enzymes. Whether or not it would destroy the minerals I have no way of finding out.

## A Homely Example

Let me give an example of the effects of an abundant feeding of the young. We bought a spotted Poland sow at a sale. She gave birth to two pigs only. We named them Theobold and Emogene. As these two pigs received a super-abundance of food they grew at such a rate as to make jack's bean stalk look like a mere sprout in comparison. Our county agent weighed them when they were six weeks of age and found that Emogene weighed 62 pounds while Theobold weighed 65. Later Emogene went to the butcher and he told me he never saw such a large hog. He did not have room to hoist it till he cut off its head. I sold Theobold to a farmer who said he had to butcher him as he was so big he could not be confined in any fence but would put his snout under the bottom wire, raise it up and the staples would come snapping out of the posts for many rods.

Now observe this: We had other pigs that had the same care after they were weaned but there were seven in the litter and they did not have the food while they were young as the two mentioned had, so the 7 were just ordinary pigs.

## Importance of Food for the Queen Larva

The importance of food for the young applies especially to the queen larva as it grows at such an amazing speed. It grows as much in proportion to its size in one day as a calf does *in a year.* At this rate if we keep the young larva away from food while grafting for 20 minutes it is the equivalent to keeping a calf away from its mother a week. We keep a few registered Guernsey cattle and the calves are allowed to run with their mothers for two months. At the present price of milk each calf consumes a hundred dollars' worth of milk but the resultant animal is well worth a hundred dollars more. Add to that the pleasure we get in rearing animals we are proud of!

## The Case of the Two Virgins

From the looks of these two virgins one might get the impression that one was wise and one was foolish similar to those ten mentioned in that yarn Mathew spin in which the ten stopped at a filling-station to get some oil. The five foolish ones were so busy gossiping about that bridegroom they hoped to see that they forgot all about getting oil while the wise ones said, "Fill 'er up." But no. The difference between these two virgins was simply a matter of diet. The small one suffered from malnutrition during infancy.

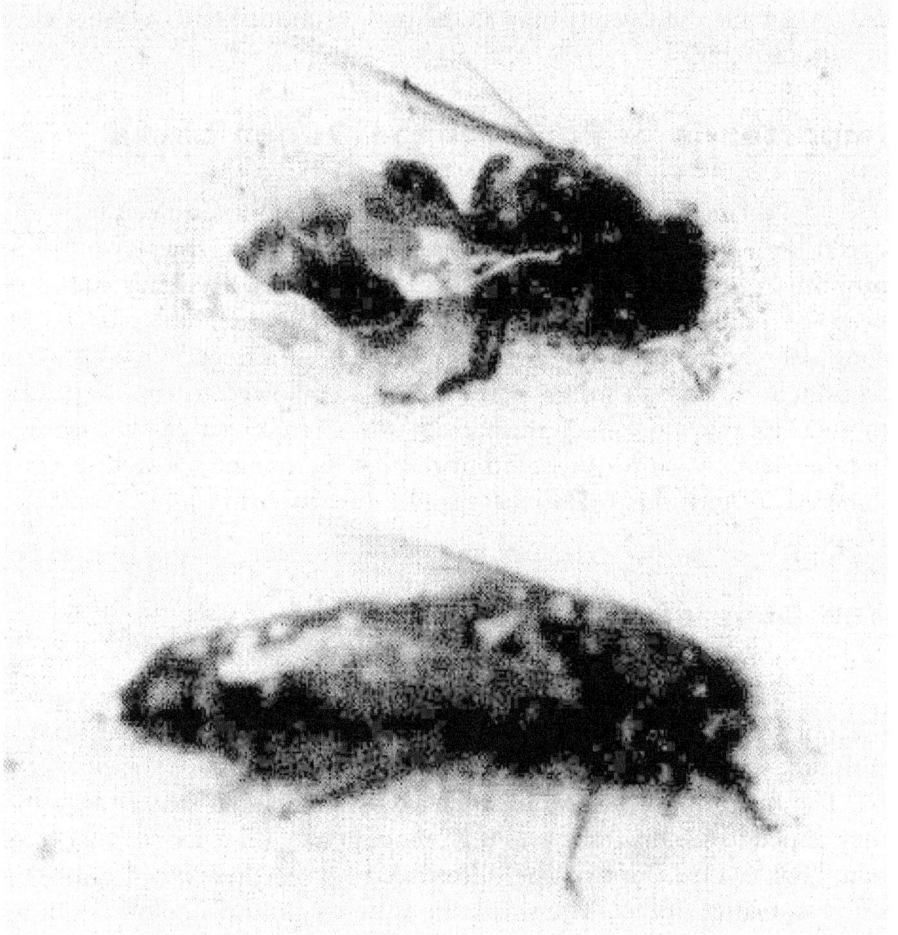

Lament of the small virgin.

You may wonder what makes me so small
Why my bees got no honey at all.
It is sad to relate but the truth I must state
'Twas because I was grafted, that's all.
The large virgin has this to say:

Why I keep my hive well filled with brood
Is because I had plenty of food,
Raised direct from the egg, the only right way

That's why my results are so good.

## An Amazing Spectacle

Having proved that perfect queens can be produced only by proper feeding of the queen larvae, I carried out an experiment to see how small a queen could be produced using too old a larva and one that had been reared in a worker cell and had been fed as a worker larva. This experiment astonished me beyond description.

The larva used was between three and four days old. A patch of such brood of about four square inches, counting both sides of the comb was given to some queenless bees. Now what did these bees do about this hopeless condition? First they chose two larvae which they attempted to make into queens. They fed them what food they had in their glands. Next, *they removed every single larva in the comb*. Why did they remove the larvae? I wish I knew the answer to that but my guess is they did it for one of three reasons and possibly for all three. First, they wanted to give all their milk to those would-be queens. Second, they wanted to use all the milk in those worker cells. Third, possibly they punctured the larvae and salvaged what milk they contained. In 24 hours they removed one of the queen larvae probably to use what little milk there was in the cell. They did their best but the cell never hatched. Therefore, when someone tells you that bees make such a mistake as using too large larvae when smaller ones are present, don't believe him. My observations have taught me that bees do the very best that can be done under existing circumstances.

## So Let Us Get to Work

I can imagine someone saying, "all right, all right, you have us convinced, so get to work and tell us how to rear those super queens that will not be prematurely superseded and will produce those bumper crops." This I shall do. First, I shall take the reader with me through the method we use in rearing queens in a commercial way and then describe methods for the back-lotter who wishes to raise a dozen or maybe fifty in a season.

# The Breeder Hive

The breeding queen hive is the crux of our system and has been developed after many years of experimenting. The difficulty has been to get the breeding queen to lay consistently in the new comb every day. At first we placed the new comb in which we wished the queen to lay between two frames containing honey and pollen. Sometimes she would lay in our new comb but all too often she would lay in the old combs and ignore our new comb. Next we made a hive with a compartment which contained the new comb only. It was only a partial success as the queen seemed not to like being caged in that way and often the queen would be lost, probably from worry or possibly the bees killed her for reasons they alone know. At last I hit upon our present system in which the queen lays in our new comb every day throughout the entire season and seems to enjoy it.

*The wood comb*

The cut shows the frame partly filled with wood, leaving a small comb five and one-half inches by nine and one half inches. These frames we shall call wood frames. It takes the queen but a short time to keep these combs filled with brood so she can devote her time to laying in our new comb. The standard size foundation will do very well but we prefer the size used in the Modified Dadant frames as it is deeper. A sheet is cut into four parts and fitted into standard frames. These frames of foundation may be drawn out in either the finishing colonies or in the front part of the breeding queen hive. When one of these new white combs is placed between the two combs partially filled with wood the queen at once begins to lay in it as the other two combs are filled with brood and the new comb is right between. This makes an inviting comb for the queen to lay in, which she does at once. A wooden partition separates the compartment containing the three combs from the rest of the hive. A cleat is nailed to the bottom of the hive below the partition. A space of an inch is left between the cleat at the bottom and the partition above. A zinc queen excluder is tacked over this opening. It is best to have the excluder at the bottom for if higher up the bees insist on filling the space with comb which they will not do when the excluder is placed near the bottom. When using the regular cover, the queen would sometimes get out of her partition due to the fact that wax or propolis would accumulate on top of the partition. To overcome this, the partition is extended three-fourths of an inch above the edge of the hive. We use two covers, one for the back part and one for the front. Then the regular telescope cover is placed on top. With this arrangement the queen never gets out of her stall. The cut will show how the hive is constructed better than a description.

*Putting new comb into the breeder hive*

We use a jumbo hive cut down to fit the standard frames. One may use the standard hive but it will be necessary to nail cleats around the bottom to make room under the frames. The entrance is about midway in front of the hive.

## Keeping the Breeder Hive Stocked with Bees

If cells are started every day the bees in the breeder hive will rapidly decrease in numbers as the only emerging brood they will have is what comes from the combs in the wood frames, therefore, some way must be provided to keep up the strength of the colony. We open the starter hive in front of the breeder hive when we take the started cells out. Many young bees will take wing so when we remove the starter hive, these young bees will join the bees in the breeder hive. This is a great help but shortly they will need more bees in which case we shake the bees from not more than one comb from the starter hive. In this way their strength will be kept up. Sometimes in the past we dumped the bees from the starter hive in front of the breeder hive. This proved unsatisfactory as so many bees were apt to kill the breeder. When they did not the breeder was apt to stop laying for a time. Adding the smaller number of bees has never given us any trouble. We prefer the perforated zinc to any other excluder. Those wood and wire excluders are an abomination. I purchased a hundred of them years ago and now all I have to show for them is a bushel of wires and slats of wood. They cost more and are worth less. The wax worms will eat through the wood and the mice will eat holes in them. Some say the bees pass through the wire excluders better. I put that question up to the bees and I quote, "There is no difference. One is just as bad as the other". End of quote

## The Feeder

This is an important part of the breeder hive. When no nectar is coming in the bees must be fed if we are to get that lavishly fed larvae we have pointed out is so necessary for best results. It is built on the back of the hive. A hole is bored in the back of the hive so the bees can get into the feeder. Slats are put in as illustrated, then melted wax is poured in to make the feeder waterproof and also to wax the slats tight to the bottom in order that no syrup or honey can get under these slats to sour. A cleat is nailed to the hive just above the feeder to keep out the rain. When feeding, the cover of the feeder is slid back and the honey or syrup poured in. As there is no bee disease within many miles of our apiary we feed off grades of honey. Honey is better food for the bees than sugar syrup but if disease is present sugar syrup should be fed. However, from years of observation I am confident the nurse bees cannot secrete milk with proper nourishment when sugar syrup is fed. I was surprised when I first put honey in the feeder. It was during a robbing season and I wanted to test it for I had a theory feeding with it would not cause robbing. It proved better than I expected for by looking at the bees at the entrance you could not tell they were being fed. With any other feeder I ever tried we had to feed at night or the robbers would rob out the colony. What causes robbing is that the bees get daubed with honey and then go out in front to get more air. The bees from the other hives pounce upon them and then proceed to clean up the whole hive. With our feeder few if any bees get daubed and those that do have to pass through a cluster of bees, then through the queen excluder and then another cluster of bees. By that time they are licked clean and do not have to go out for an airing.

## Stocking the Breeder Hive

In case your breeding queen is in a fairly strong colony, the bees and brood are placed in the front compartment and one frame of brood and the queen is put in the back compartment. The two frames fitted with wood are placed in the back compartment with the frame of brood between them. Feed is then placed in the feeder. As soon as the queen has eggs and brood in the two outside combs, the frame of brood is removed and a new white comb is placed between the two combs. In case the colony is strong, most of the brood may remain in the front compartment, but later most of this must be removed to make room for the combs containing eggs. As stated, we should have several drawn combs ahead. When the brood in the front compartment is removed frames with sheets of foundation may be given. As the bees are being fed, they draw the foundation readily. The comb given should be left 24 hours when it will be well filled with eggs. It is then placed in the front compartment and another frame of new comb is given. This is repeated every day until there are three frames containing eggs in the front compartment. When the fourth comb of eggs is put in the front, the first one given will now contain tiny larvae just out of the egg and as this is practically all the brood that large numbers of bees have, the tiny larvae will be found to be fairly flooded with bee milk. In giving comb to the breeding queen, it should be trimmed back to the foundation provided the bees have built comb at the edges and bottom of the foundation. The queen prefers to lay in natural built combs rather than those built from foundation. I have used all kinds of foundation and find this to be true. The comb shown was not trimmed back so it can be seen that the queen laid in a sort of horseshoe shape.

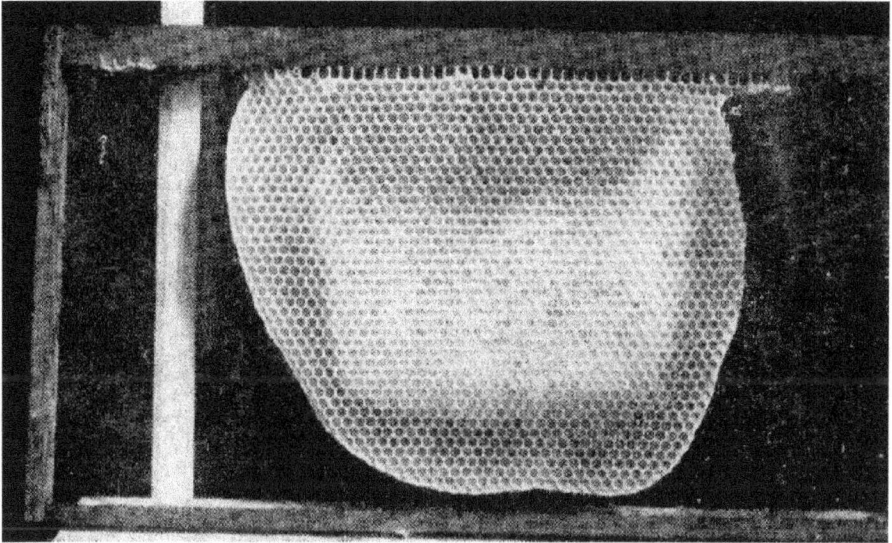

*Larvae ready for use in cell starting*

The bees do not build cells with such regularity as the cells in foundation, therefore such comb is difficult to cut as many larvae will be destroyed. When the eggs are laid in cells drawn from foundation it is much better if the row of cells is uniform and easily cut in a straight line. Now let us examine the larvae in the cells to see if they are getting lavish feeding, the necessity of which we have so frequently stressed. The larvae are so small they can hardly be seen with the naked eye. They are flooded with bee milk fully as much as larvae in swarming cells. In these same cells they will remain while the bees go right on feeding them. Let us now prepare to start those larvae on the road to queendom.

## The Starter Hive

*The starter hive*

    The hive in which we start cells we call the starter hive. In the years that have passed since I began rearing queens in 1901 I have done a vast amount of experimenting. I used a starter hive when I first tried my hand at queen rearing. Then I would think I had a better way and would abandon it. Later I would come back to it, then leave it again for something I thought better. With some improvement in it we are now back to it to stay.

    Many years ago Henry Alley used what he termed "the swarming box". He did not start cells in it but confined bees in it for some hours and at night dumped them into a hive with no brood or queen. He seemed to think it necessary to keep the bees queenless for some time to get them to accept the cells. Then the prepared cells were given to the bees in a hive containing two frames of honey and pollen. Mr. G.M. Doolittle said the bees would accept cells better if kept queenless for three days. In my experience I find the bees accept the cells just as readily after an hour or two as they do if left queenless for a longer period. We usually keep them confined in the starter hive from one to two hours which gives them plenty of time to clean up any honey that may have been spilled on them when shaking them into the starter hive. In later years Eugene Pratt gave the cells direct to the bees confined in a box and he called it "The swarm box". We prefer to call it "The Starter Hive" as it is used for starting cells and has no relation to a swarm. Our starter hive is made wide enough to hold five standard frames. The total depth is 17 inches. It has a wooden bottom with cleats at the sides making it suitable for feeding or watering bees. Three frames of honey and pollen are put in *but no brood*. In case no nectar is coming in, the bees that are to be used in stocking the starter hive should be fed three days before using them and this feeding should be continued each day until they are shaken into the starter hive.

## Food and Water Necessary to the Bees in the Starter Hive

A matter that I believe most of us have overlooked is the necessity of food and water for the confined bees. All have observed how bees carry water in abundance when rearing brood. Now this water is even more necessary when rearing queens for the bees require more water when producing the milk in abundance. The best way to provide this is to fill the center comb with diluted honey. This not only provides the necessary water but the food it contains acts as a flow of nectar. In case some 50 cells are to be started from five to six pounds of bees should be used. In case from 75 to 100 cells are to be started from 8 to 10 pounds of bees should be put in. If larger number of bees are put in the starter hive should be kept in the shade or some cool place. The screens on the sides provide an abundance of ventilation.

# Getting the Bees with Which to Stock the Starter Hive

In getting the bees for the starter hive we used to hunt up each queen before shaking the bees. This entailed a lot of work and sometimes when it was cloudy it was well-nigh impossible to find the queens. Again when robbers were troublesome, unless we found the queen quickly, we had to close the hive and go to another. In case we used the bees from three colonies, at the best it took at least half an hour and usually longer. With our present system the starter hive may be stocked with bees from three or four colonies *in five minutes*, and this too, rain or shine, robbers or no robbers.

The colonies are all in two stories with queen excluders between the two hives. The queens are confined below and the top story is kept well filled with brood. All that is necessary to do in filling the starter hive is simply to shake in the bees from the top story. The bees will be depleted somewhat, as some are used to keep up the strength of the breeder hive and the finishers, so unsealed brood should be given to these colonies from other colonies. We use the bees from each colony once in every four days. This is less work than it would be to use a larger number.

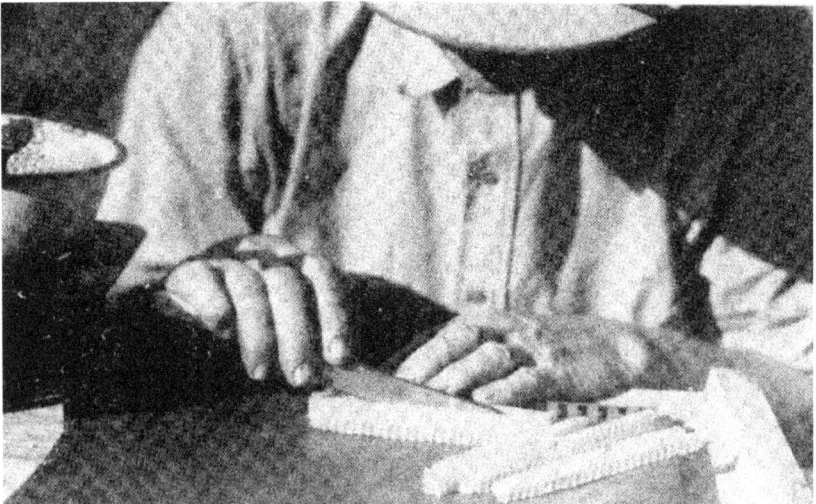

*Cutting the comb into strips*

# Preparing the Cells

Let us now consider the larvae in the breeder hive is ready for use. First some wax should be melted. An electric hot plate or a sterno stove with canned heat may be used.

*The cooling box*

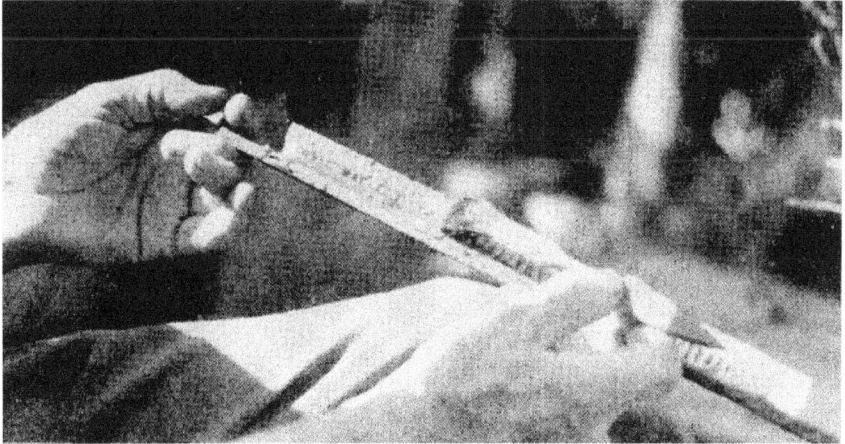

*Painting the cells*

*Destroying the larvae*

It will be necessary to destroy some of the larvae in the cells in order to make room for the queen cells. We have tried many methods but find the best way to destroy the larvae is to insert a spike in the cells and punch out the bottoms. An ice pick may be used if the point is broken off. We usually destroy two cells and leave one which gives plenty of room for the fine large queen cells. Sometimes when we are short of cells we destroy every other one. Some will be built close together and will be destroyed in cutting but one can get more cells in this way.

The comb from the breeder hive is now cut into strips one cell wide. With a small paint brush, wax is painted on the cell bar and the strips of cells placed on the waxed bar. Then with the brush wax is painted on the sides of the cell strips. The cells are then cooled in the wooden tank and more wax painted on till both sides of the strip of cells have enough wax to make the cells strong enough so they will not mash down when inserting the spike. Wax should be cool for painting the cells, nearly ready to solidify. We use a wooden tank four inches wide inside measurement, and half an inch longer than the cell bars. A cleat is tacked in one end under which the end of the cell bar is placed. At the other end of the tank is a nail driven through from the side extending about two inches into the box. As soon as the cells are waxed the bar is shoved under the wooden cleat at one end of the tank and the other end is placed under the nail at the other end. Then cold water is poured in the tank till it covers the melted wax. It is well to put ice cubes in the water for we want the cells to be firm so the larvae in the cells can be destroyed without mashing down the comb. While we prepare the cells as quickly as possible in order that they may be in care of the bees without delay, it has been astonishing to note what gluttons for punishment these well-fed larvae are. Frequently we have more cells than we have use for so instead of using the larvae, we set the comb aside for the larvae to die in order that we may use the comb again. Often after four days, when we give the comb to the bees in the breeder hive, some of the larvae will be rejuvenated or resuscitated, or whatever you wish to call it. An interesting feature is that although the larvae were away from the bees four days, they had not increased in size a single bit. This shows that larvae away from bees do not grow, so we should keep them away from the bees as short a time as possible.

## Putting the Prepared Cells Into the Starter Hive

We use two bars in each frame, the space above and below the bars being filled out with wood. The starter hive is given a little jar on the ground so the bees will fall to the bottom and not fly out. The two frames of cells are set in place, the comb containing thin honey being in the middle. Cells will be built better in the middle of the combs as shown, for where the cells are near the top of the frame the bees are apt to build comb over them.

## Bees in Perfect Condition for Starting Cells

Let us now consider the bees in the starter hive. They have been nursing a horde of larvae in the colonies from which they were taken. They are secreting an abundance of bee milk. They are now away from their brood. They feel their queenlessness. The milk in their glands seems to cause them discomfiture. They want to rear queens and they want to get rid of that surplus of milk. When giving them the bars of cells they at once set up a joyful humming to spread the glad news that their wants are fulfilled.

Often when I come with the prepared cells they set up a humming when the bars of cells are within a foot of the starter hive. Evidently they smell the larvae in the cells I am giving them. Since some who read this may not know, I will state that at the tip of the abdomen of the worker is a scent gland. By exposing this gland and fanning their wings, the bees spread the scent to the other bees. Mr. Eugene Pratt is accredited with the discovery of this gland in the year 1891. By the way Mr. Pratt gave us something to consider when he wrote, "Sugar, bees and brains are the prime factors in queen rearing." I have often stated that many of us fail because we use too large a proportion of bees and sugar!

## Program of These Rejoicing Bees

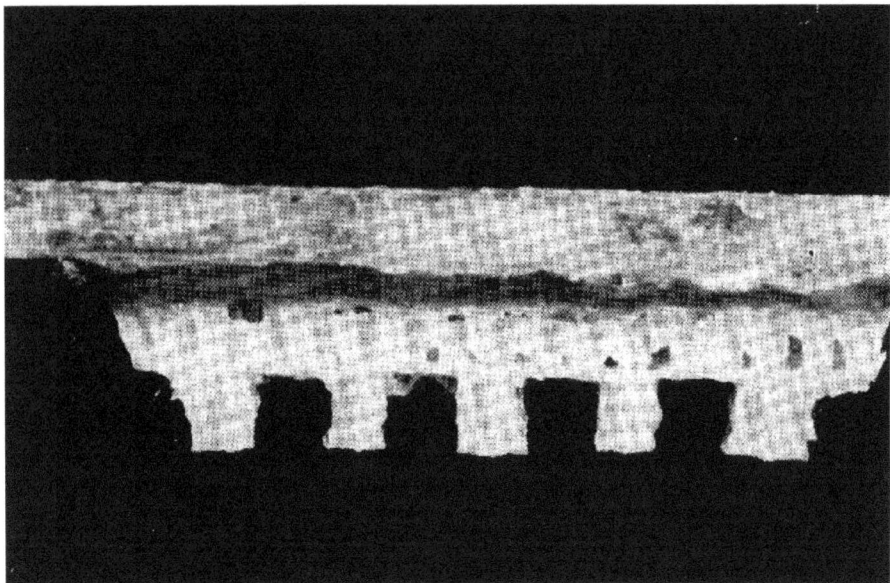

*Cut #1 shows the cells as we prepared them especially for this picture. We bent the sides of the cells over to destroy the larvae in them leaving one cell and destroying two.*

Now what is the first thing the bees do upon receiving these prepared cells? They first unload the accumulated milk in their glands by feeding the larvae. *Next, they begin tearing down the walls of the worker cells.* It has been stated by many that the bees enlarge the worker cells. This they cannot do. They tear down the worker cells and build queen cells over the larvae. In this way they build the cells as they want them and do not have to put up with any man-made contraptions. The photographs will show how bees build their cells.

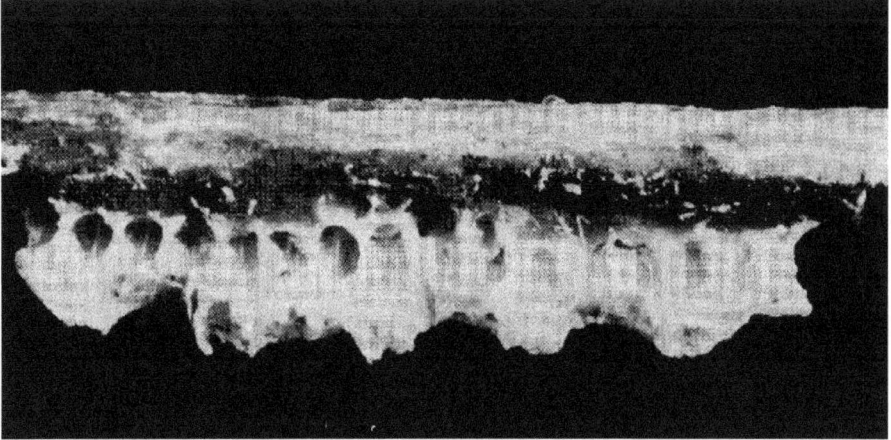

*Cut#2 shows how the bees have torn down the worker cells and have started to rebuild. One cell at the right has already begun to take shape. This photo was taken only four hours after the cells were given to the bees. I had intended to take more pictures four hours later but darkness come on, so number three was taken the next morning and, as can be seen, the bees are rushing the work along at top speed. Note what a fine start they have and yet the larvae in these cells are no larger than those usually used for grafting.*

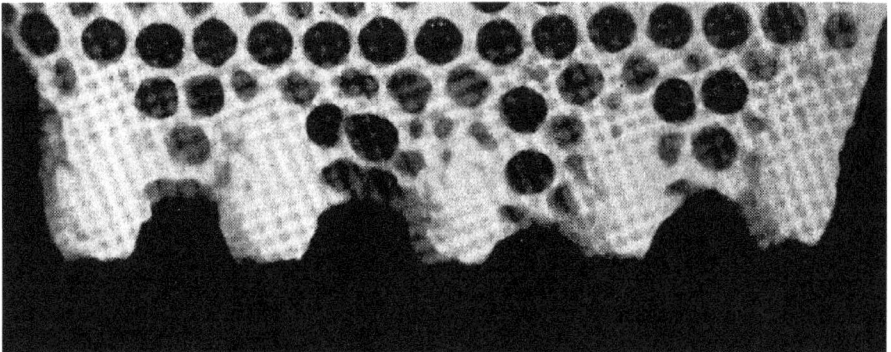

*The next picture was taken 24 hours after the last one and now the bees have fully half an inch of bee milk stored in each cell.*

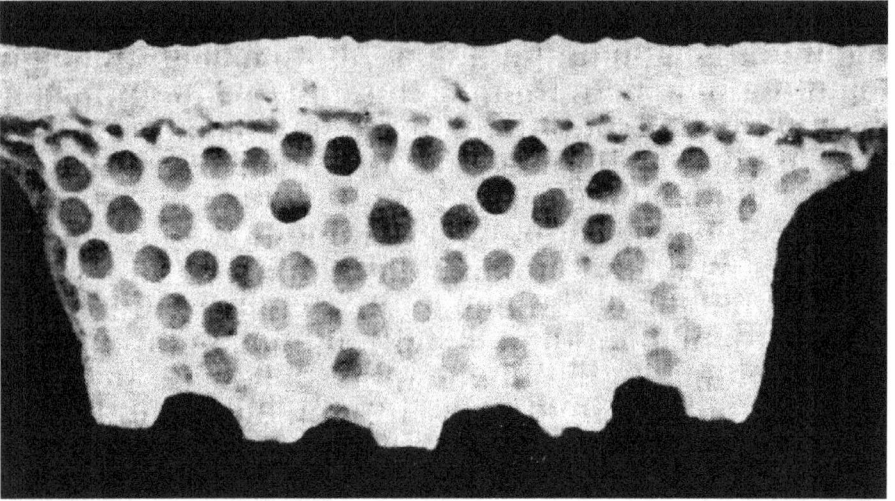

*The fifth picture was taken 24 hours after the previous one and, as may be seen all the cells are capped over except the one at the right and it was capped over two hours later.*

When starting cells from worker cells the bees first "raze" them and then "raise" them. This reminds me of an incident that happened years ago. I told my wife that a virgin had gotten into the hive where I had a lot of queen cells and this virgin was razing cells. She thought I said the virgin was "raising hell" so I let it go at that for I rather liked her version of the incident.

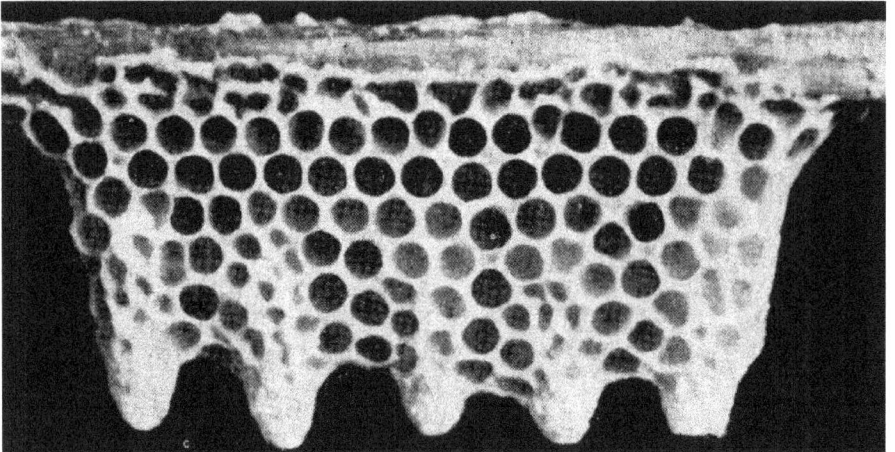

# Finishing the Cells

While the bees are now busily engaged in remodeling the cells, we must prepare to have them finished, for they should not remain in the starter hive more than 24 hours for the best results. Again I shall tell what we do and then tell why we do it. The best colony for finishing cells *is a strong queenless colony.* The cut shows cells made by the grafting system illustrated in that excellent little book by Walter Kelly, *How to Grow Queens for Fifteen Cents.* Also is shown the use of wooden cell cups.

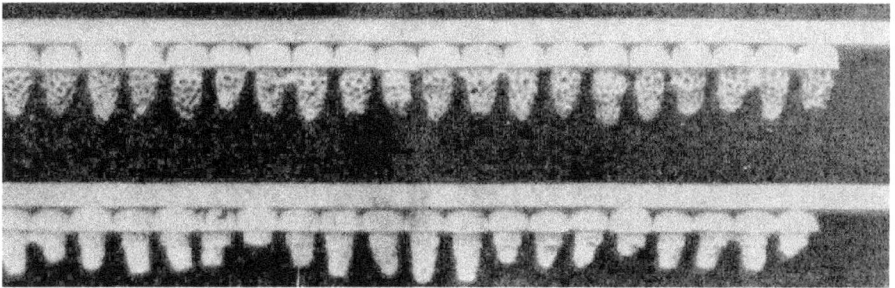

Thus far our system is foolproof as far as getting good cells started, for if we give them too many cells for best results, they will accept only the number which they can properly build. Right here is an interesting feature of bee behavior. While using the grafting method, the bees would accept all the cells we gave them, while in reality they were in no condition to properly care for one-fourth of that number. Why they accept more artificial queen cells than they can properly care for is due to a peculiar characteristic of the honey bee in that they will finish, or try to finish, any job that has been started, while had it been left to them in the first place they never would have started it. This trait may readily be observed if you place brood in a hive above a queen excluder with the queen below. The bees seldom start queen cells from this brood. But if cells be started and placed above the excluder the bees will finish them the best they can. Therefore, when artificial queen cells are given above an excluder, they evidently consider them as started cells so they do the best they can to finish them. When cells are given them as we do, they evidently do not consider them as started, therefore they accept only the number that they can complete in a perfect manner. Many times in Indiana we were trying to rear queens when conditions were not the best. Our customers were demanding their queens or their money back so we were trying to raise queens when conditions made it impossible to rear queens in quantities. Many of these cells, when the larvae were three days old, had little bee milk around the, barely enough to keep them alive. I knew from experience that only poor queens would result from such stunted cells and have taken my knife and shaved off hundreds of them and destroyed them. Had we been using our present system, the bees would not have accepted more than ten or a dozen in place of a hundred but would have finished them in perfect manner. To be extra conservative in the starter hive not more than half the number of cells they would accept. Therefore, as stated, thus far our system is foolproof as far as getting good cells started but from now on the beekeeper must take over, for as the cells now have been started the bees you give them to will try to finish them whether or not they can do a good job at it. Unless the bees that are to finish them are in the best of condition and strong in numbers you can still produce poor cells. The queenless colony that is to finish the cells must contain at least five pounds of bees and several combs of mostly capped brood. If no nectar is coming in, this colony must be fed liberally. Again, to be extra conservative, we have two colonies to finish the cells that have been started. If this is properly done you can produce cells every bit as good as those built by the bees during swarming. Bees must be added to keep up their strength. It is some help to them by giving them the cells with the bees that are with the cells. When more bees are needed the starter hive is dumped in front of the finishing hive.

## Nurse Bees

Let us digress for a moment and discuss nurse bees. Some who have made a study of the subject tell us bees are good nurses at three or four days of age. Well if they have caught them in the act and really have the goods on them I will have to retract, but even so maintain that, if they are nurses at that tender age, they are might poor ones. I asked one professor, who claimed these young bees were good nurses, how he came to that conclusion. He said he had observed the young bees putting their tongues into cells where the young larvae were and he supposed they were feeding them. I asked him if he should see a boy crawl out of bed at ten o'clock at night, climb up into a chair and put his hand into the jelly closet he would deduct that the boy was putting the jelly there? Now, when these young bees put their tongues into the cells, do they give food or take it? In my experience I believe it is the latter. Acting up such conclusion I tried a plan in which I expected to get wonderful results. I shook bees from a dozen colonies in front of a hive in which I expected to rear cells. As I wished, the old bees went back to their own hives while the young ones went into the hive prepared for them. There were at least six pounds of those downy, fuzzy bees. They had been fed well so I gave them 50 grafted cells expecting to get wonderful results as I had been told these young bees were graduate nurses. Upon examining them next day not a single cell was accepted. All those little beggars did was just *eat*. That is what they are supposed to do, eat and develop and grow in order that they may be able to do any work about the hive and they can then do any work that is assigned to them such as nursing, gathering nectar and pollen, carry water or guard the entrance. I have found a bee worthless as far as doing any useful work about the hive until it is a week or more old. When ten days old it can do anything required.

## Old Bees Good Nurses

Again some have told me that old bees cannot nurse as their milk glands dry up. I maintain such a statement erroneous. Many have observed that bees returning from the fields could not nurse. Of course not, for they were acting as fielders and not prepared to nurse. I will wager (a dime is my limit) that if these same bees were given several frames of unsealed brood, they would soon change from fielders and become nurses. Some of the brood might perish, for this was thrust upon them with no warning. But they would at once put on the white aprons and white caps with red crosses on them, take a hurried course in nursing, and in two or three days could show those younger bloods how to nurse. In a few days should you remove all the brood and give them grafted cells, they would do excellent work. Now just what are the requirements when taking this course in nursing? They devour pollen and honey in abundance and soon begin giving milk.

I have given a frame containing eggs only to a colony that had been queenless and broodless for some time. The bees removed many of the eggs as they were not nurses. Then they concentrated on a few and built a few very small queen cells. The eggs were given to them on too short notice for them to become nurses. Had they been given three day's notice they would have taken a course in nursing and would become competent nurses.

I have taken queenless and broodless colonies that had been in that condition so long that I was afraid laying workers would develop, and have introduced a laying queen. The presence of this laying queen served notice on them that they must take a course in nursing. This they did by eating pollen as has been stated and in three days, when the eggs began to hatch, they fairly flooded the cells with bee milk to such an extent that one might believe they were going to make queens out of the whole lot!

Probably in a normal colony the bees do the work best fitted to their age but, as stated, they can do any work required after they are ten days old. In the far North in the spring many bees would be six months old and all would be at least four months old. These old bees do a better job of nursing than the young ones that come latter, for European Foul Brood seldom attacks the first cycle of brood, while later, after the old bees are gone and the young nurses take over, European Foul Brood develops. This would prove that the old bees are better nurses.

## Finishing the Cells-Continued

We use a single story to finish the cells unless there is a heavy honey flow on, in which case we add a second hive body but have the cells finished in the bottom hive. If we start four bars of cells in the starter hive it is best to give them to two finishing colonies. As the bees finish a job that has been started we must be very particular to see to it that they are in the best possible condition to finish the cells properly. In case we wish to start cells every day we have three sets of finishers, two hives in each set. In case one is rearing queens on a very extensive scale he can have four hives or more in each group and use two or more starter hives. After the cells have been in the starter hive overnight we lift out the frames of cells and give one to each of the finishers. The adhering bees are given to the finishers which helps keep up their strength. The frame of brood and cells are arranged as follows: At the side of the hive nearest to you is a frame of capped brood, then a frame of cells. Next day, group number two are given bars of cells, then next day group number three are given two bars of cells. Now, when you start the fourth batch of cells, the ones in the first group are sealed or nearly so and are given bars of cells, a frame of capped brood separating them from the first frames given. This is repeated till all finishers have three frames of cells each. When ready to give the next frames of cells, those first given now are what is known as "ripe cells." They must be taken out to make room for the next frames, and are ready to be introduced to mating hives. These cells will hatch the following day.

The photograph on page 60 shows the arrangement of frames of cells as they are painted white that they may be easily seen. From this on, when a frame of ripe cells is taken out another frame with started cells is given and this may be continued throughout the season. As the bees in the finishers become weak more bees must be added from the starter hive and capped brood given from other colonies. In case you do not need so many cells you can skip a day and the space in the finisher is filled by moving the combs over and should there be a honey flow an extra comb should be given to prevent the bees from building combs in the space.

*Three frames of cells*

## Caution

Three or four days after giving brood you should look over the frames to remove any cells that may have been started. It is astonishing to note how many accidents can happen in those finishers no matter how careful one may be. We tack a zinc queen excluder to the bottom of the finishers to keep out virgins; for if there is any place one can get in is sure to do so. These queenless bees frequently send out calls for a queen by fanning their wings so it is necessary to make the hives bee-tight. In one finisher I found a laying queen and removed her. A week later I found another in the same hive. Upon carefully examining the hive I found a small opening near the top of the hive where the cover did not fit closely and the queens returning from their mating flight were invited in. By making the hives bee-tight and by examining the brood carefully you will have no accidents, or rather I should say, few.

## Reviewing the Past

Before giving the ripe cells to the bees in the mating hives, let us go back into the past and examine the various methods we have used in starting and finishing cells. I would prefer to omit this but some may think they have a better way and may try some of the plans I have tried and found wanting in one respect or another. I believe I have tried every plan that has any merit and many that have none. As stated, we used a starter hive in our first experiments in 1904. I used wooden cell cups and the grafting method. Just why anyone should use wooden cell cups is now a mystery to me. Possibly for the same reason I did-because I was taught to use them. Later we dipped the cells as Doolittle did and this saved a lot of time as we did not have to clean out the wooden cups and insert the cells into them. Later, I put the wooden cells into the furnace and they made a fine fire and kept the house warm for half an hour which was of more value than using them. I also bought a lot of queen nursery cages for the cells to hatch in and some of them did. Many did not and I raised a lot of half starved virgins that never amounted to anything.

## Requirements of the Newly-Hatched Virgin

When the virgin emerges she is weak and frail like most newly-born creatures and needs the very best of care. She must be fed in the natural way by the bees until she is mature which is from 8 to ten days. To confine her in a cage and expect her to develop into a good queen is to expect the impossible. We found when a virgin was caged for three days she usually was delayed in mating three days so nothing was saved in time but much was lost in the quality of the queen. Some queen breeders have advocated keeping the virgin in the cage for five days. Such queens are not worth introducing. It would be better to allow the bees to supersede their queens in the natural way than to introduce such worthless queens. To be sure, there may be more money in it for the producer of such queens in case he could continue to find buyers but the buyers certainly would not make a profit in the deal. So my cages went into the furnace and did more good there than they did in hindering the development of the virgins. I wish to take just one more dig at another abomination and then we will get on to constructive work. That abomination is the wire queen cell protector. I understand that this bright (?) idea was conceived by two men about the same time and they came near developing a regular Kentucky feud as to who deserved the credit for inventing it. They might just as well have remained good friends for really there was nothing worth quarreling over. In using these cell protectors one is fighting bee nature and in such a fight the bee wins. The idea of the cell protector is that as bees tear down the cells at the side and not the ends, by protecting the sides the bees cannot tear them down so the virgins are allowed to emerge. Again we are trying to *force* the bees to accept cells they do not want for if they could get to them they would tear them down. So what are the bees going to do about it? They do just this about it. Figuratively speaking, they just sit down in their easy chairs and keep an eye on that cell and when that impudent baby dares to poke its head out of that cell they pounce upon it and tear it to bits. They simply will not tolerate having that unwanted baby planted on their door step. Once more we are attempting to force the bees to do something against their will.

## The Remedy

What can we do about it? Do not attempt to force the bees to do your bidding but use appeasement. Feed the bees well and get them in that frame of mind such as you have after a big Thanksgiving dinner of roast turkey stuffed with oysters served with cranberry sauce and then topped off with hot apple pie with ice cream. After such a meal you "jest hain't mad at nobody." Well fed bees that have been queenless a day or two consider it a favor to have a ripe cell given to them. They accept it with thanks and at once begin to fasten it to the comb, air-condition it and eagerly await the birth of the baby, and when that happy event occurs they nurse it with loving care and give it a good start towards becoming a perfect queen. At any rate I had about a bushel of these protectors. What grieved me was that they were noncombustible. I often called them worse words than that. Since they would not burn I had to bury them, and as I remember there were no mourners at the grave and what I said would never qualify as a prayer.

# Early Experience in Cell Production

I first used a starter hive with holes in the lid in which were placed flanged cups. A cushion was placed on top to keep the cells warm. The theory was that in giving the grafted cells to the bees as soon as they were grafted one could get better results. Lots of work and poor results were my reward. In order to get better results I put more bees into the starter hive. The bees would fret and if the weather were warm many would suffocate. Then the starter hive was abandoned and another plan tried. Two hives were set to the rear and a single hive placed where they formerly stood. Two frames of honey and pollen were placed in the empty hive. About half of the bees from the two hives were shaken into the hive that was to receive the cells. This gave very good results but I thought there must be a better way, for this system had two disadvantages. First as the colonies were deprived of their bees they ran down rapidly. The other objection was that often we wanted a larger number of cells than the two colonies were able to start and we had to prepare another set which made the labor excessive.

## A Colony for Both Starting and Finishing

One season we used queenless colonies for both starting and finishing cells. Cells were given every four days. Before giving more cells, more bees to the amount of three or four pounds were shaken in from other colonies. These colonies did not accept the cells as readily as we wished but finished very well what they accepted. The bees never fanned and showed they were delighted to receive the new cells. The presence of other cells may have been a contributing cause. To sum it up, the results were not the best and the labor was excessive.

## Three-Compartment Hive for Both Starting and Finishing

Another plan suggested itself. A three-compartment hive was made in which cells were both started and finished. The two outside compartments contained laying queens. These were separated from the central compartment by queen excluders. The bees from both sides flew from the central compartment. When giving cells, the openings between the compartments were closed and bees from both sides were shaken into the central compartment. Flight holes were then opened in the two outside compartments. Bees flying from these openings would join the bees in the central compartment. The plan promised much but proved to be about the flattest failure of anything I ever tried. They accepted very few cells and those they did accept were not finished well. So the hive was junked. I did not burn it in the furnace as I did the wooden cups-here in Florida we have a fireplace. I then realized what a stupid idea it was after all. Why have the two hives connected? Why not shake as many bees as you needed into the starter hive?

## Starting and Finishing in the Same Hive Not Practical

In reviewing all the plans I have tried in the past I came to the conclusion that no plan is practical in which cells are both started and finished in the same hive. One can shake bees into a hive containing two frames of honey and pollen and allow them to both start and finish the cells. This may be a very good plan for the small beekeeper who wants but a few cells. However, this would not be practical for producing cells in quantities as it would take a large number of colonies and they would be greatly weakened.

## Back to the Starter Hive

Once more I returned to the starter hive but built along different lines and it has given such perfect results that I am sure we will never use any other method for starting cells. When everything is considered it has more advantages and fewer disadvantages than any other system. Cut out of this starter hive has already been shown.

## Queenless Colonies Best for Finishing Cells

As stated, we finish the started cells in queenless colonies. I finished cells above the excluder with a laying queen below for thirty-three years before I realized this is a very poor place to finish cells. Let us examine some of its shortcomings. In the first place the top hive is where the bees like to put their honey. During a honey flow they fairly bury these cells in honey. This smothers many of the larvae and the bees tear down the cells. Those not destroyed are smeared all over with honey making it a messy job in handling them. All of this could be overlooked if we were getting the kind of cells we wished, which we were not. The main objection is that since they have a laying queen below, they have about all the brood they can properly nurse, therefore they have neither the ability nor urge to build good cells.

## Ability and Urge Necessary for Producing the Best Cells

Before going any further, let us look into this matter of "urge." Before bees can do the best at cell building they must have both ability and urge to do so. A weak, poorly-fed colony deprived of its queen would have the urge to build cells but would be lacking in ability. On the other hand, a strong colony headed by a young queen all in a three story hive packed with brood would have the ability to build good cells but would be lacking in urge to do so. To remove all their brood and queen would give them both ability and urge to build the very best of cells. We start cells in the starter hive because it gives us the very best results. The bees shaken into the starter hive have been nursing larvae, so by taking away these nurses and putting a large number of them into the starter hive they have ability and urge in the highest degree. After the cells are started, they may be given to a colony having less urge and all will be finished well provided the finishing colony be strong and well fed. When these nurses have been taken away from larvae they have been nursing they act as if in distress. Possibly the excess milk in the glands in their heads gives them a headache. The best headache remedy is to give them larvae to nurse. When the cells containing larvae are given to the bees confined in the starter hive, they at once begin fanning in order to spread the glad news and at once unload the milk accumulated in their glands.

## Bees Having Access to Their Queen Lacking in Urge

As long as bees have access to their queen they will not do the best at cell building. When a frame of larvae is placed above an excluder with a queen below, the bees seldom start cells. With the laying queen below, the bees have about all the larvae they can care for, so they have little urge to build cells. It is plain then that finishing cells above the excluder with a queen below is not the best method when quality of queens is desired.

Another proof I have observed *thousands of times* is found in the hive containing the breeding queen. As before explained, the frame of eggs is placed on the opposite side of the excluder from the queen. I have yet to see a single queen started there even though there is plenty of larvae just the right age. However, should the queen be killed, or if I remove her to introduce another, thus making them queenless, the bees at once fairly cover this comb with queen cells. I wish to give just one more example where there is lack of urge to build cells when a queen is present, although in this case she did not even have access to the combs in the hive. My son Huber and I were requeening some colonies of hybrid bees we had acquired. One colony had a young laying queen and there was an abundance of brood in three stories. There was plenty of pollen and there was a light honey flow on. Hunt as we would we just could not find that black queen. As we had a queen in an introducing cage and as it was getting late, we had to do something about it. We knew we were using unorthodox methods but as an emergency existed, we thought the bees might overlook this little irregularity and accept the queen. We therefore set the hives aside, put a queen excluder on the bottom board, shook all the bees from the three hives on the ground, replaced the hives and put the introducing cage in the top hive. As the hive stand was high we doubted if the queen could find her way up to the entrance of the hive and if she did she could not get in.

We thought those bees up in the top story would be delighted to get our queen. We just *thought* that. The bees thought differently about the matter and killed our queen as soon as she was out of the cage. Now note this interesting fact-not a single queen cell had been started although it had been six days that queen had been below the excluder. We found this diabolical black queen on the bottom board and killed her and the bees accepted our next queen with thanks.

## The Question of Mating Hives

Before discussing mating hives I will tell a story and it is a true one. A number of years ago a man came to see me who was conductor on a passenger train. He had held that position for 25 years and had laid up some money. He said, "I think I have served the people long enough as I have worked hard as conductor. I have about decided to buy a little farm and about 500 colonies of bees and just take life easy and let the bees work for me. What do you think of it?" I replied, "As conductor you just think you've been working. Now if you want to know what work really is you get 500 colonies, do all the work yourself, then you will realize that as conductor you just *thought* you were working." He insisted there could not be much work caring for bees. All there was to it, you just put on the supers and when filled take them off and people would come running for the honey. He may have been reading an advertisement of a supply manufacturer that ran something like this, "Costs little to start. Practically all profit. No experience needed. Very little work," etc. What I am driving at is that most of us are like that conductor more or less and I hope less.

*The miniature mating hive*

That is the attitude many have in relation to mating hives. Many think that most any kind of box made of thin material will do. Nothing is farther from the truth, for the size and the way it is made is very important for best results. I have in the past, in an experimental way, made 16 different styles of mating hives. I have made them in sizes as small as the one shown in the cut, which used a single individual comb honey section about one inch square, and from that up to a two-story jumbo hive. In Indiana we used a mating hive that held three four-by-five comb honey sections with a division board feeder at the back. It did very well in Indiana but was a complete flop here in Florida as I rather expected it would be. Here is the general rule: The smaller the mating hive, the less number of bees it requires and the less feed it takes to operate it. However it requires much more labor to operate it as constant care must be taken to see that it does not run short of stores and to see that it does not become too weak in bees or that it does not become too crowded with bees or stores, for if any of these conditions are present, the bees will abscond.

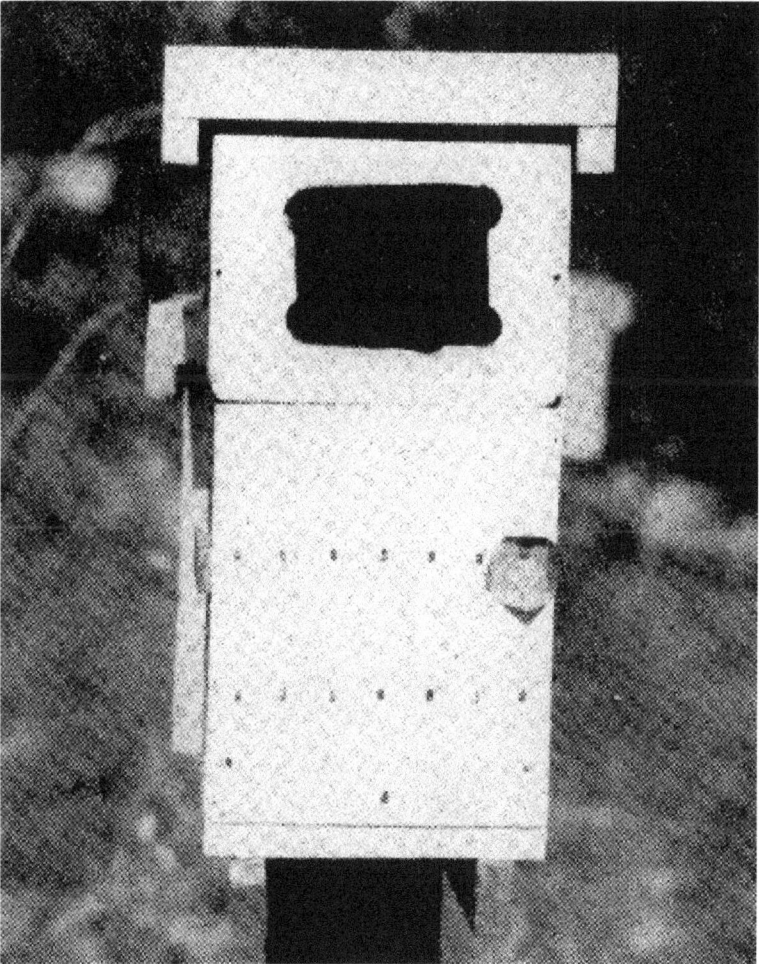

*The Adolescent hive with ventilated box on top*

# The Adolescent Hive

The most successful small hive I ever used is what I term my adolescent hive. It is larger than the baby hive. It is made to hold three frames 5 x 8 inches. It is made of full three-fourths inch material and the cover is made of 2-inch material.

A small mating hive should not have a ventilator, as the air will blow in through it and cause the bees to realize their weakness and they will abscond. For an entrance a half-inch hole is made in the center of the hive. This hole is slanted upward to keep out the rain. If the entrance is placed at the bottom, the ants or robber bees will kill out the bees if they get weak.

*Our five frame mating hive*

When feeding, a battery filler can be used and a little food may be squirted into the entrance at night. A package box which is used for bringing bees from the outyard is made with wire screen sides as shown in the cut. Half a pound of bees will nicely stock this hive. A little feed is given, a ripe cell is placed between the combs, and the package box is placed on top. The entrance should be kept closed for 24 hours if the bees are taken from the outyard. If taken from the home yard the bees should be confined 48 hours, or many will return to their former colonies.

After confining them for the required time, the entrance is opened and the package box removed. To those who prefer a small hive this is by far the best of anything I ever tried. Even this hive requires close attention to see that it is kept in the right condition. We have wintered queens in such hives here in Florida but the results were rather uncertain so we have adopted a larger mating hive, one holding five standard frames. These hives are placed on concrete standards, two hives on each standard. The covers are made of one-and-one-half inch material. This cover keeps them warm in the winter and cool in the summer. They are heavy enough so no wind will blow them off. At first we formed them in spring. This weakened the colonies in the outyard to such an extent that they gathered no honey. Now we form the mating hives in the fall after the honey flow is over. We proceed as follows: As the bees in the outyard must be requeened, we take a package of one and one-half pounds of bees and the old queen. The bees and queen are put into the mating hive. As they have a queen it is not necessary to confine them if they are put in after dark. A young queen is introduced into the colony from which the package was taken. With a young queen, the colony soon makes up for the loss of the bees. After the old queen in the mating hive has been laying for a week or so she is removed and a ripe cell given. With this management, the colonies in the outyard are actually benefited. As we winter the colonies in the mating hives there will not be many to establish in the fall, only the ones that have run down for one reason or another. Some mating hive colonies may be too weak for wintering. In such cases a package with queen is put in the same as though the hive was empty. When there is no honey flow, our five-frame hive requires more feed than the smaller one. On the other hand the bees in our larger hive will gather more honey and therefore need less feed. While rearing queens we have but four frames in the hive as it is much more convenient to handle the frames. After queen rearing is over in the fall, an extra frame of honey is put in. A few may die out in the winter for one reason or another, but the rest will be so strong that they may be divided to make up the loss. When introducing cells, when no honey is coming in, we feed the night before giving the cells. Should any be short of stores, they are given an extra amount of feed. After these hives are nailed up we boil them in used crank-case oil. When dipped in cold oil, they will not take paint but when boiled in oil the oil soaks in and later they can be painted. We do not know how long such hives will last but as the oil penetrates the wood, and if we keep them well painted I believe they would last a hundred years. The entrance to this hive is in the center of the front. It is one and one half inches in diameter and the hole is slanted upward to keep out the rain.

## Bees Abscond More in Florida

Why bees abscond more in Florida than in Indiana is not clear. It was stated they abscond because the ants bother them. We made the standards ant proof but it made no difference. Bees abscond here from the baby mating hives when they have virgin queens, something we never knew to occur in Indiana. Therefore, the beekeeper must study conditions in his locality and act accordingly. As Joseph C. Lincoln would have one of his characters say, "If you got intellects, use 'em." My old friend, Allen Latham, uses a baby mating hive covered with tar paper and in his cool New England climate it gives him good results. Should his system be tried here in Florida, the bees would last till the sun got a good chance at the hives, when bees, comb and honey would come running out at the entrance. Our hives are placed on concrete standards so they come about waist high. The labor in handling them is much reduced with this arrangement. How anyone can tolerate having the hives low down on the ground is a mystery. We had ours that way in Indiana and my back still aches from it! Help will do a third more work with hives on standards.

## Introducing the Cells

In introducing cells no protectors are used but we simply press the cells lightly against the combs, placing them near brood as the bees will care for them better. A certain high brow told me cells do not hatch as only eggs can do that. Instead we should say the virgin emerges. From a grammatical standpoint he is correct but when I get ready to say anything I do not propose to let grammatical expressions or rhetorical technicalities hamper the thought I wish to express so, Mr. Highbrow, we shall continue to "hatch" our cells.

## Stocking the Mating Hives

*Huber using the bee hopper*

In stocking the mating hives it is best to take them from an outyard as they do not have to be confined so long. We do not weigh the bees but measure them instead. We have what we call our bee hopper. It is four and one half inches wide. The bottom is covered with galvanized iron. A scoop is made just wide enough to reach across the inside of the hopper. For the five frame hives we use about a pound and a half of bees.

## Why Bees Tear Down Cells

Well-fed bees very seldom tear down cells. Just why I do not claim to know. I discovered this fact about thirty years ago. We were then using a mating hive holding two frames of honey. The bees tore down the cells just as fast as we gave them. I had been told that bees tear down cells because they are strange to them, similar to their reaction when introducing a strange queen. If that were true it seemed that little could be done about it. I sat down on a stump to think it over for most things can be worked out if we use the right formula. I remember that bees tear down cells worse at times than others. Why? It could not be because the cells were strange to the bees for they were always strange yet at times few were torn down while at others, as at present, nearly all were destroyed. Another reason was this: In introducing cells I often had some left over and would put them back into any hive I was using for cell building, often putting them into a hive that had not built them, yet in doing this many times, *there never was a single cell torn down*. That was proof that the bees did not tear them down because they were strange. Then why did the bees in the mating hives destroy the cells while the bees in the cell builders did not? Plainly it was because the bees in the cell-building colonies have been well fed while the bees in the mating hives had not. To test out my theory I fed the bees in the mating hives that were to receive the cells the next day. The more I thought of it the more certain I was that it would work. *But*, what would the bees think about it? Cells were given the next day and I was delighted to see a high percentage of them was accepted.

## Well-Fed Bees Do Not Tear Down Cells

So my conclusion was that well-fed bees do not tear down cells. True, even during a heavy honey flow some cells will be destroyed but in such cases there may be a large amount of unsealed brood and few fielders so in reality the bees are not well fed. Liberal feeding will make acceptance sure. When there is a honey flow we do not feed but remove the laying queen one day and give a cell the next day and have very little loss. Often, when we had an abundance of ripe cells, we removed the layer and at the same time put in a ripe cell. When there was a honey flow we had a good acceptance but the results are rather uncertain so we do not recommend it.

## How Long Should the Bees be Confined to the Mating Hive?

In case bees are taken from the outyard they should be confined to the mating hive one day and released after dark the day following. Some may ask why confine them at all when taken from the outyard? If liberated at once they do not seem to realize they have been moved and at once fly out without marking their hives with the result that many will gang up and go into one hive leaving the hive from which they came depleted in bees. This I have experienced in the past to my sorrow. At one time we took a hundred baby hives stocked with bees to an outyard and at once liberated them. Immediately the air was full of bees and after circling about for some time they all tried to get into one baby hive and they covered it with bees about a foot thick! We surely had a time trying to distribute them. When confined for a day and released at night, they seem to forget their old home and behave as well-bred bees should. In case bees are taken from hives in the home yard close to where they are to be established they should be confined two days and released after dark. Most of them will stay but a few will be gifted with remarkable memory and will go back, but there will not be enough to materially weaken the bees in the mating hives.

## Mating Hive Records

One must have some system of knowing what the conditions are within the mating hives. We have used a number of systems and out of all of them we have worked out a system that gives such perfect satisfaction that we never expect to make a change. We need to know the conditions within the mating hive for a period of two weeks, for by that time the virgin should have been mated and by that time she is a laying queen. Therefore, we have a row of copper tacks across one side of the hive, 14 in number, one for each day of the week for two weeks. These tacks are driven about halfway in. The first tack at the left represents Monday. By putting a mark over Thursday's and Sunday's tacks, we can see at a glance what days we are working on. Tags are made of galvanized iron two inches square. A square hole is made in the center of the tag. The corners are bent forward a little for easy handling. They are first dipped in yellow quick-drying enamel. After drying, each corner is dipped in enamel of the following colors. First black, then red then white, then blue reading to the left. Now let us consider the hive and it's signal relation to the queen. Let us say we introduced the first cell on Monday. The tag is placed on the tack at the extreme left and turned to red, indicating a cell has been given. We examine the hive Wednesday and if the virgin is there the tag is turned to white. Should the virgin not be there, the tag is turned back to black indicating the hive is queenless. When another cell is given, after the first one is not accepted, it is put in on Wednesday so the tag is moved to the third position and turned to red. This process is continued throughout the season. A picture of our mating hive has already been given. To indicate the condition within the hive we have blocks four inches long painted black so they may be seen at a distance. A number of these blocks are distributed among the hives, being placed on the stands between the hives. In case the hive can spare honey the block is placed lengthwise of the hive at the extreme left. If it needs honey it is placed crosswise in the same position. The block shown in the cut shows the hive can spare both brood and honey. If placed crosswise at the extreme right it indicates that the hive needs brood. If placed lengthwise on the same position it indicates the hive can spare brood.

## Introducing Queens

Since the year 1900 I have spent much time in experimenting with introducing cages and hundreds of queens have been sacrificed in the interest of science or the lack of it as you will. I have tried all the fads such as daubing the queen with honey, sprinkling peppermint water on them, strangling the poor long-suffering queens with tobacco smoke and have tried many other such crack-pot methods with dire results. Some still just daub the queen with honey and drop her in. When you do that I suggest you say, "And may the Lord have mercy on your soul." Many queens "introduced" in that way will be killed outright but what is even worse, many will be injured and a queen may remain at the head of the colony some time before being superseded thereby losing a crop of honey. I have seen many queens that have been maltreated and then allowed to head a colony. One queen in particular was shiny with no hairs left and was completely devoid of clothing which might cause one to infer that she either was a member of a nudist colony or had lost in a game of strip-poker. In my years of experimenting I have come to the conclusion that the bees and the queen must become acquainted and no short cut can accomplish this. During a honey flow one may have a certain degree of success with any of these "quick" methods which will continue till disaster overtakes him. One party wrote us that he had never lost a queen when using the regular mailing cage so we could not interest him in any better method. Later he wrote that in introducing 12 queens he had lost five and wanted to know what was the matter. I wrote him that there was nothing the matter as that was perfectly normal. Probably he had formerly introduced his queens during a honey flow and the latter when there was no flow. Another party drove about 200 miles to our apiary to get five queens as he said he did not want them to go through the mail as they might get injured. As he seemed so careful I offered to lend him some of our introducing cages. He declined to accept our offer stating he just dropped the queens in honey and then put them right in among the bees. I told him I would not give a nickel a piece for queens introduced that way. Later he wrote that he lost two in introducing and wanted me to replace them. No, we did not replace them but did give him a lecture on introduction. I doubt if it took, for we never heard from him again. I have known queens that were not properly introduced to stay on the bottom board for several days before beginning to lay. Then they would lay for a short time and be superseded. I do not know what caused the injury as they did not appear to have been balled. The queen being a very nervous creature may have received a nervous shock which injured her. We must remember it is as much the attitude of the queen as it is of the bees that makes for perfect introduction. When the queen emerges from the regular mailing cage she is often frightened for she suddenly finds herself among strange bees. Often in her fright she goes taxiing over the combs squealing as she goes. As she does this the bees seem not to know just what is going on but realize there is some excitement and they want in on it so, as the queen goes by, a bee may grab a leg and others join in and form a ball around the queen either killing or injuring her. Sometimes they sting her but not often. Sometimes the queen's legs are broken. Such queens never amount to anything and the owner of such queens will report that he does not like that strain of bees for they just do not get honey.

# How Soon May a Queen Be Released Among the Bees?

Our vast amount of experimenting along that line leads us to believe that when conditions are not favorable, as when there is no honey flow, the queens should not be liberated under *four days*. When conditions are favorable, as when there is a honey flow, three days will suffice. Even then, if released from an ordinary mailing cage, she may be injured. Now, how does our cage remedy all these defects? In our cage we use the principle that is acknowledged by all to be good. This principle is described in the bee literature of 60 years ago. That principle is allowing the bees to get into the cage with the queen instead of the queen getting out among the bees. As far as getting frightened is concerned, the case is reversed, for when a worker gets into the cage with the queen, it is the worker that is frightened. In my experiments, I have often observed the behavior of the first bee that gets into the cage with the queen. The first bee to get into the cage often would buzz around trying to get out. After a time it would flatten itself out and lay on the bottom of the cage, stretch out its tongue and offer food to the queen. She is too well bred to refuse so they become friends and band together for mutual protection. Soon another bee enters and it seems more frightened than the first for now there are two bees for it to contend with. However, they soon become acquainted and more bees come in and the queen begins to lay, the bees either eating or destroying the eggs. In another day the bees eat out the candy at the end of the partition and the queen goes out among the bees, not as a fugitive but as an accepted queen, and soon will have a large patch of comb filled with eggs.

## Requirements of the Perfect Cage

First, the cage must be made of wood. A metal cage will burn the queen in the summer and chill her when the weather is cool. We sacrificed a number of queens by using a cage containing too much metal. We made a few cages of cork but the bees chewed up the cork in short order.

The cut shows the best cage we ever made, and by its use we have never lost a queen or had one injured due to the fault of the cage. We lose a few queens due to our own fault such as introducing a queen into a colony that is superseding its queen. We remove one queen, not knowing there are two, and in such a case our queen is killed as soon as it is out of the cage.

Even at that our losses will not run over two percent and none whatever when everything is normal. Our cage has two compartments, one for the queen and the other for the candy. I have found that the bees eat just three-quarters of an inch of candy every 24 hours. This they do with surprising regularity.

The drawings of our introducing cage were furnished through the courtesy of *the Southern Beekeeper*. The drawings illustrate the cage better than photographs.

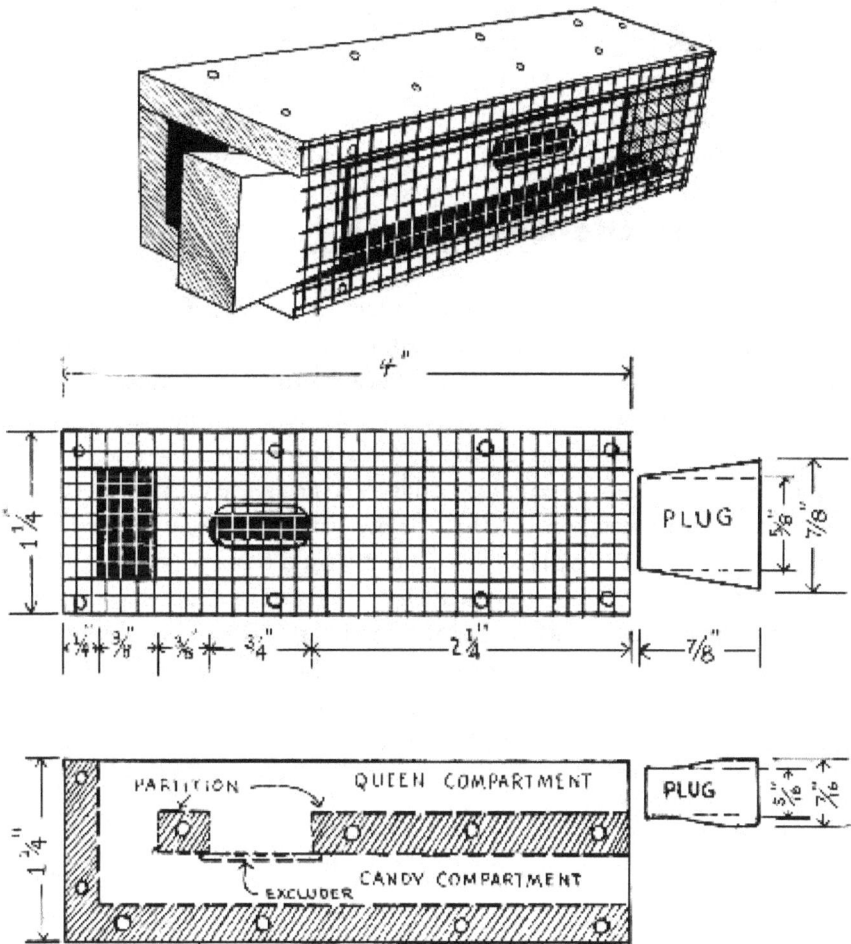

In the partition between the two compartments is a piece of zinc queen excluder. This is just two and one-fourth inches from the opening of the cage. The partition extends to within a quarter of an inch of the back of the cage. The compartment on the opposite side from the screen is filled with bee candy.

The queen is placed in the second compartment with no other bees. It takes the bees just three days to eat out the candy to the queen excluder. After the bees enter the compartment with the queen they eat from both ends of the candy, so they eat out the candy in one day and release the queen. When there is no flow, the candy compartment is filled completely so it is four days before the queen is released. When there is a flow, the candy compartment is filled to three-fourths of an inch of the opening, thus releasing the queen in three days. This cage is not patented, so anyone handy with tools can make it but it is best to have a sample.

## The Push-in-the-Comb Cage

Years ago I brought out a cage with a wooden frame with metal teeth for pressing into the comb in order that the queen could lay in the comb before being released. When properly handled, this cage was sure in its results but too often it was not properly handled. It had to be used on old combs or it would fall from the comb and the queen would be killed. Then the hive had to be opened in order to remove the cork to let the bees into the cage and it was not suitable for outyard work. Again we could not use the automatic eat-out plan as the queen would remain on the comb and refuse to come out. Altogether, with the average beekeeper, it was not a complete success and we do not recommend it, nor do we now use it, as our present cage is just as sure in its results and the beginner can get just as good results as the experienced beekeeper. In using our cage, two combs may be spread apart and the cage placed between them. It may be placed in the super or on the bottom bar of the brood frame. The cage should never be placed on the bottom board as the ants may get in and kill the queen or she may be chilled.

# Does Opening the Hive Cause the Bees to Kill Their Queen?

My answer to that is positively no! Often we open the hive as soon as the queen is out of the cage and in thousands of such cases we never had a queen killed or injured. A common belief is that the hive should not be opened until two weeks after the queen is introduced. In case the queen is not properly introduced, the bees will ball her whether or not the hive is opened.

## The Home Yard Cage

We have what we call the home yard cage-the cut will show how it is made.

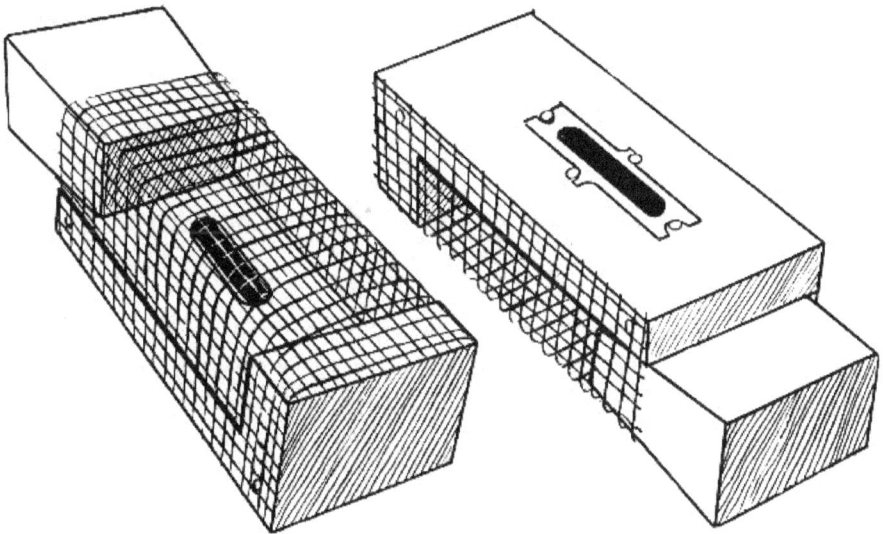

Often we have one or two queens to introduce in the home yard, and do not take the trouble to load a cage with candy, so we use our home yard cage. In the wooden side is a queen excluder. The queen is put into the cage and the cage is laid on the top bar or bottom bar with the queen excluder down so the bees cannot get in. In two or three days, depending on conditions as previously mentioned, the cage is turned over thus allowing the bees to enter. In one day more the queen is released. This is just as sure as the candy cage but requires more attention to opening the hive and turning the cage over at the right time.

## Bee Candy

Candy for the introducing cage or for mailing cages is made by mixing XXXX powdered sugar with honey. This should be kneaded as stiff as possible. If the honey is heated it will absorb more sugar. If one has bee disease, or is unable to get a health certificate before shipping, candy may be made of invert sugar and such candy is nearly as good as that made with honey. To make invert sugar, put half a pound of water in a kettle and let it come to a boil. Then add one pound of granulated sugar and stir in one-fourth teaspoon of tartaric acid and boil till a temperature of 248 degrees is reached. A candy thermometer is necessary for this. No certificate of health is required by the postal authorities when using invert sugar candy. All that is necessary is a statement that there is no honey in the cage. After the invert sugar is cooled it is used the same as honey in making bee candy.

## Shipping Queens

Our experience has taught us that the reason so many queens die in the mail is lack of ventilation or because they are overheated. We have found that by wrapping the cages in cloth, they go through in good condition if they are not too long on the road. We also ship in mailing tubes which give even better ventilation. Shipments made with mailing tubes have gone to 43 foreign countries by air with almost no loss.

## A Review

As we thought it necessary to do a lot of explaining as we went along, we will now give the procedure step by step:

1. Stock the breeder hive as described, putting queen and one frame of brood and the two wood-comb frames in the back compartment, putting the frame of brood in the center.

2. Prepare as many colonies as are needed by putting them in two story hives with brood above the queen excluder and the queen below.

3. Cut Modified Dadant foundation in four pieces and put them in frames and get this foundation drawn in any colony suitable.

4. Take out the frame of brood from the back compartment of the breeder hive and put it on the other side of the partition, being sure the queen is in the back compartment. Then put in a frame of the drawn foundation. Leave it there for 24 hours, then put it on the other side of the partition. In case you wish to start cells every day, put in another frame of drawn foundation in the back compartment. Repeat this every day until the fourth comb is placed the other side of the partition, when the first is ready for use.

5. Put three frames of honey and pollen in the starter hive and put some thin syrup or diluted honey in the middle comb. Then shake in bees in the amount of five or six pounds from the hives previously prepared. Leave them three or four hours.

6. Take out frame of young larvae and cut into strips, mount them on bars and give them to the bees in the starter hive.

7. Next day remove the queen from two colonies to be used as cell finishers and give to each colony one frame of cells and dump the starter hive in front of the colonies from which the bees were taken.

8. When all finishing colonies have three frames of cells each, and the fourth is ready, remove the first given and introduce the cells to mating hives.

## Starting Cells Every Third Day

In starting cells every third day, but two finishing colonies will be needed. In this case you put in a new comb in the breeder hive every third day and when it is placed the other side of the partition, an old comb or a frame of brood is placed in the back compartment between the two wood-combs. There are both advantages and disadvantages in starting cells every third day instead of every day. We are running more to the third-day plan. One advantage it has is that the drawn sheet of foundation may be left for the queen to lay in longer. Another is that as cells are introduced every third day, in introducing the cells, the ones introduced three days before can be examined. If using the third-day system, in order to keep the record straight, one should write down on a piece of paper the letters C S I, C S I in rotation, these letters being placed at the left hand of the paper, one below the other down the paper. C stands for comb and indicates that you gave a new comb to the queen in the breeder hive. S means that you start cells, and I means you introduce cells. Put the date and the day of the week after each letter. Of course in starting there will be no cells to start or introduce, but after some time the letters will have a meaning. In case one started cells regularly every third day there would be little use for the above memorandum as the condition of the cells in the finishing colonies would indicate what is to come next. However, in case you slipped one starting, all cells would look much alike and you might get things mixed up. At any rate I have done so and when I came to giving the started cells to the finishers I found the hive full of virgins!

## Modifications for the Small Beekeeper

For the commercial queen breeder, or for the honey producer who wishes to rear 200 queens or more I strongly recommend the management and equipment that have been described. For the small beekeeper, or the back-lotter who wishes to rear a dozen or maybe 50 queens in a year, no special equipment will be necessary. Under such circumstances it is best to rear queens during a honey flow as conditions are more favorable and queen rearing is a simple matter compared to rearing them when there is no flow. When the honey flow is just beginning is the best time to start as there is an abundance of pollen coming in along with the nectar, and pollen is just as necessary as nectar.

For starting cells, first get the foundation drawn as described. Next set to the rear the hive containing your breeder, with the entrance in the opposite direction from the original position. In its place put another hive and in it put two frames solid with honey and pollen. Shake in half of the bees from the breeder hive and put the breeding queen into the prepared hive. Within 24 hours she will have a good patch of eggs in the new comb. The comb should be placed over an excluder on a strong colony. In case you do not wish to start more cells immediately, this hive with the two combs and the one with the eggs may be set over the hive formerly containing the breeder, setting this hive on its former stand. The two hives must be separated with an excluder, with the queen in the bottom hive. This colony must be well fed for we want the larvae that will soon come from the eggs to be well fed. As soon as the eggs hatch, turn a strong colony to the rear and in its place set a hive into which you put two frames of honey and pollen then shake out half of the bees from the hive you have moved into the hive with the two combs. In case the colony is not of sufficient strength, two colonies may be set to the rear and half of the bees from both colonies shaken in. Next cut the comb into strips as has been described and give a frame of two bars of prepared cells for this colony to both start and finish. In case one prefers to use a starter hive, one can be made by merely tacking a screen to the bottom of a hive. If this is done the hive should be set on blocks high enough to afford plenty of ventilation.

## Mating Hives for the Small Beekeeper

In case increase is desired a hive may be moved to a new location and another hive set in its place. Into this hive put two or three frames of brood with adhering bees from the hive moved, being careful not to get the queen. It is well to wait two or three days after dividing this colony before giving a cell for bees will keep returning from the hive moved and are apt to destroy the cell given. As soon as the queen has mated and laying, the hive should be filled out with combs or frames of foundation or starters. In case feeding is necessary a good way to feed is to tip the hive back a little and under the front place a block. Then at night pour in a little syrup. Be careful not to put in more feed than the bees will consume during the night for if any syrup is left it will incite robbers as a bottom board feeder is the worst of any feeder to cause robbing.

## Mating Queens in an Upper Story

In case one does not wish to make increase, a good way to mate queens is in an upper story of a two-story hive. A frame covered with wire screen should separate the two hives. Always use a wire screen and *never a queen excluder*. Some in the past have advocated using a queen excluder but in the great majority of cases the bees will kill the virgin as soon as she emerges. In the frame of the wire screen an opening of half an inch should be made for a flight hole. This entrance should be to the rear of the hive below or the queen in returning from her mating flight is apt to get into the hive below.

*Hive with rear entrance*

The entrance to the top hive should be closed for two days, otherwise too many bees will join the colony below. Syrup or honey should be put into one of the combs and a ripe cell given. After two days the entrance should be opened and the virgin will soon fly out and mate. When the honey flow permits, this is an excellent way to requeen the colony for we have *two laying queens* instead of having it queenless while introducing a queen or cell. The bees in the top hive will not suffocate from having the entrance closed for the screen will afford abundant ventilation. The best way to put syrup into the combs is with the use of a pepper box feeder. That is a mason jar with the cap full of holes. With this the syrup may be shaken into the combs. If you merely want to requeen, all that is necessary is to hunt up the old queen below and remove her and take out the screen. The bees will not fight as they have kept up their acquaintance through the screen. In case increase is desired, just move the bottom hive to a new location and set the top hive on the old stand. The lower hive should be moved as it is much stronger than the top hive. The bees for a time will try to get into the back of the hive but after a time they will make a chain around to the front door and all will be well.

## Requeening by Giving a Ripe Cell

As a rule we do not recommend giving a ripe cell to a full colony, as it weakens the colony by being deprived of a laying queen for nearly two weeks. Again, a strong colony does not accept a cell as readily as a weaker one, and if the cell is not accepted it will further weaken the colony. However, when the honey flow is over and more bees are not needed immediately, it is often desirable to requeen by giving a ripe cell. As it is difficult to find a virgin in a strong colony, one may give a cell and without looking for the virgin another cell is given three days later. In a week after the first cell is given, the colony may be examined and if there are no queen cells started, you may know a virgin is present. In case both cells are destroyed, or should the virgin be lost in mating, it is best to introduce a laying queen as the colony would become too weak if given another cell.

## Clipping the Queen's Wings

The question has often been asked, "Does it injure the queen to clip her wings?" In case the wings are clipped too close to the body it may injure the queen as veins and nerves are cut. We clip the tip ends of both wings which prevents the queen from flying and in no way injures her. We do not advocate clipping one wing as the queen is apt to be injured. We used to clip one wing but too often just as we squeezed the scissors the queen would poke a leg between the blades with the result that she got an amputation along with the clipping. We prefer to have all the queens clipped in the outyard that we may know if any have been superseded.

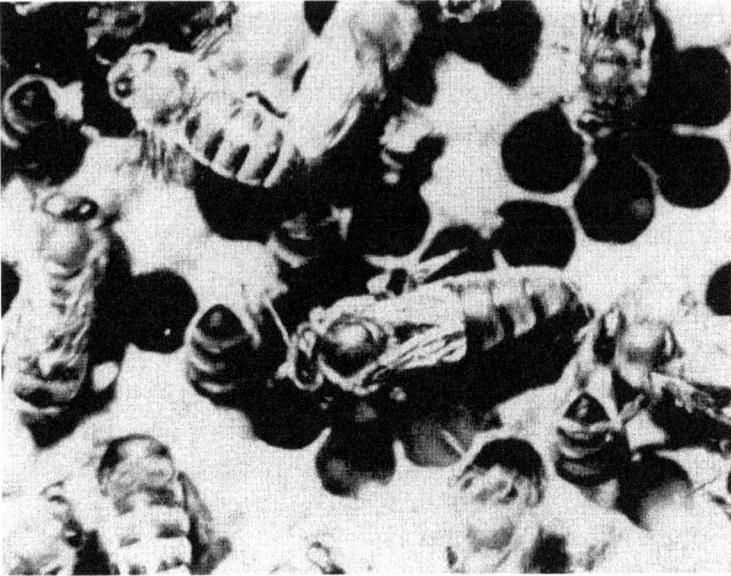

## Do Queen Cells Interfere with Introduction?

As far as we have been able to determine, the bees accept a queen when queen cells are present just as readily as when there are none. However, one must be sure the cells are not too far advanced. Cells that have just been capped or are about to be capped we pay no attention to but if they have the appearance of being well along and nearly ready to hatch, we go over the combs carefully and remove every cell. If this is not done a virgin might emerge before the laying queen is out of the cage and of course would kill the layer, for a fight between a virgin and a layer always results in the death of the layer. Even though the layer is out of the cage and laying, she might not get around to destroy the cells and a virgin would emerge and kill the layer for her lack of good judgment. A laying queen is often very reluctant to tear down cells as in nature she is very seldom called upon to do this, while a virgin is a pastmaster at this kind of sabotage for I have known a virgin to tear down over a hundred cells in 24 hours.

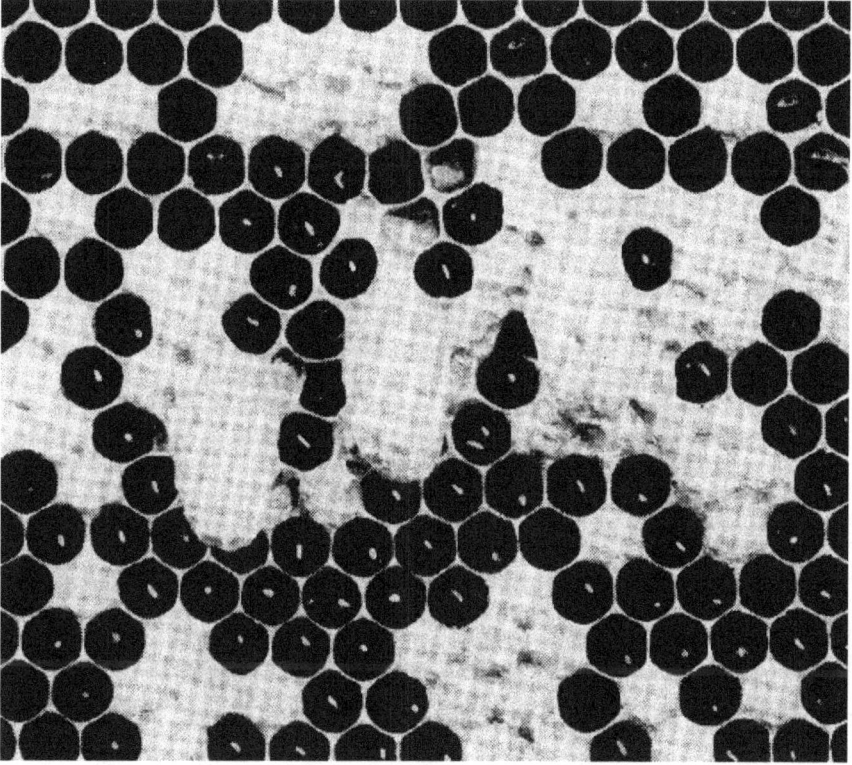

*Eggs around the queen cells*

The photo will show a case in point. A laying queen had been introduced and she seemed to be either a Jehovah's Witness or a Conscientious Objector for she refused to do battle with those cells. I photographed this and wanted to see the negative before destroying the cells and left them in till the negative was developed. Upon examining them I found a cell had hatched and a virgin was loose in the hive with my layer. She was too young to do battle but was in training and in a few hours more would have purged that layer. I killed the virgin but that layer had a close call. It would have served her right for being such a rank pacifist.

## Breeding Queens

One must use his best judgment in selecting a breeding queen. Many have advocated breeding from the queen whose colony made the most honey. That would be a very good plan if one could be certain the larger yield was due to the better gather qualities of the bees. Many factors enter in that make for a better yield. For instance, when bees are out for a flight for the first time the air may be filled with them. In returning to the hive many of these bees are unable to locate their own hive among so many all set close together with the result that, when they alight at the entrance of any hive, they fan their wings, sending out the scent and thereby attracting in all the bees in the air. In this way several pounds of bees may be added to a colony which results in a better yield although the queen may be an inferior one.

## The Colony that Gave a Bumper Crop

In Indiana we had an outyard laid out in the form of a triangle as that was the shape of the plot on which we had our bees. During the sweet clover flow one colony produced three supers of honey while the others averaged about two supers. In the fall that colony produced two supers of honey from smartweed and asters while the rest produced a little less than one super. Surely that colony that so far outdistanced the others must have a queen that would make an excellent breeder. I thought I would take a look at her but alas, when I opened the hive, I found it not only had no queen but was *fairly lousy with laying workers!* Just why then the big yield? This colony was located at the point of the triangle to the west and the fields of nectar lay to the west. It was evident that the bees in returning from the fields-maybe the ones out for their first load-stopped at the first hive they came to and kept it packed with bees.

## Large Bees Better Gatherers

One of the most scientific experiments ever carried on in my opinion was conducted by Doctor Merrell at the Kansas experiment station. After some exact research work in finding that some bees actually gather more honey than others, upon examining them he found that they were *larger.* That is as we would expect. Which can haul the most goods in a day, a large semi-trailer truck or a pick-up? With this in view we have tried to choose for our breeders the ones whose workers are larger. We feel sure that if any progress is made in the future in producing bees that get more honey it must be by selecting the *largest drones.* Drones vary much more in size and shape and color than do either the workers or the queens so I believe we should proceed along this line.

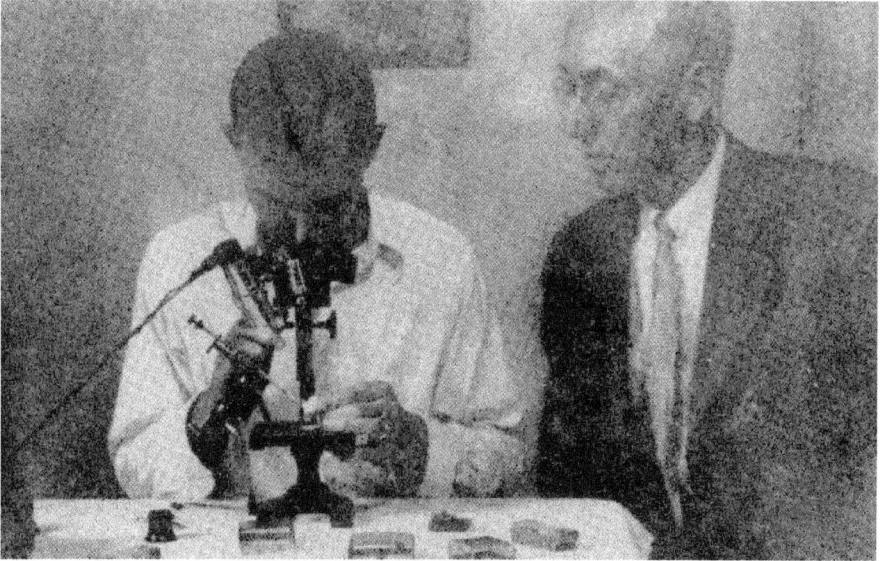

*Dr. Lloyd Watson instructs Jay Alfred in artificial insemination*

## Artificial Insemination

For several years we used the system developed by Dr. Lloyd Watson for inseminating queens. The system is a success and much credit should be given to Dr. Watson for perfecting it. However, in our case we could not afford to carry it on as it required most of the time of a skilled operator. My son Jay Alfred did this work and accomplished a great deal for the short time he had it in hand.

We are still interested in controlled mating, for no great advances can be made without it. Here in Florida we tried island mating one season, the island being located two miles from our apiary. To test the location to see that no drones were within mating distance, we took six mating hives with virgin queens but *no drones*. The idea was that if all six of these virgins failed to mate it would be proof that there were no drones within mating distance. In that case we would then weigh out our largest drones, take them there and get controlled mating. But it was not to be. Four out of six mated with the drones of our yard two miles away. We did not believe they would mate with drones at that distance, so we learned that much.

Many beekeepers claim they get pure mating as there are no bees other than their own within mating distance. I believe if they would try it as we did they would find they would get mating from drones far away. Nature is very solicitous about preserving the species and the queen and drones do court on the sly in a manner unknown to those who try to snoop into their private affairs. Possibly the antenna of the drone picks up the high-pitches sound of the queen's wings miles away. Who knows?

## Bees and Chess

I love the game of chess possibly from my nature of tackling something I can't do. I am not much of a player but probably get more of a kick out of it than the masters do. I have often said that chess and bees are similar in a number of ways. First, the queen is the most important individual in both games. If one loses a queen in a colony of bees he may requeen from the larva in the lowly worker cell. If he loses his queen in a game of chess he may requeen from the lowly pawn, or rather *maybe* he can.

Again both games are so deep that the master minds of the past have not been able to delve to one-tenth of the depth of either game. It has been stated that if all the games of chess that can be played were numbered it would represent a string of figures a yard long. So also the bee problem is so deep it cannot be fathomed. And last, the similarity of both games lies in the fine class of good sports that plays both games. As Will Rogers said about congressmen, "they are the finest bunch of men money can buy." The friends we make at bee meetings and the friends we make by letter in all lands make life worth living. Let me quote from that prince of men, W.Z. Hutchinson, in his book *Advanced Beekeeping*: "fortunately, however, the perfection of man's happiness bears but little relation to the size of his fortune and many a man with the sound of bees over his head, finds happiness, deeper and sweeter than ever comes to the merchant prince with his cares and his thousands."

## Amusing Happenings

Many amusing incidents occur while working with bees, two of which I wish to relate. These two incidents I wrote up for the journals some years ago. How often I have told them I do not know, but my wife says that if she had a dollar for every time she has heard me tell these stories she would endow a home for the feeble-minded. I do not believe there is anything personal in that remark.

## The Tramp and the Bees

During the first world war I had some bees on a lot near the back door of a house where lived a dear old lady, Mrs. Mechlin by name. She was one of those kindhearted women who could easily be imposed on. I was teaching in the high school at the time and was out before breakfast working with the bees to get as much done as possible before school time. I was meditating over the fact that help was so scarce as so many of the boys had gone away to war. Apples were rotting on the ground because there was no one to pick them up. Sweet potatoes were rotting in the ground because there was no one to dig them, and other foodstuffs were in the same bad way. While thus thinking over these things, along came a big fat tramp who went to the back door of the widow Mechlin's house and put up the usual pitiful tale, "can you give a poor fellow a bite to eat? I haven't had a bite since yesterday." Neither had I, but I was working early and late trying to do my bit while that worthless no-account tramp was begging from that dear old lady. Mrs. Mechlin relied, "Why most certainly, my good man. You just sit right down in this chair and I will prepare you a good hot breakfast." I was furious. How I wished I could get him out among those bees. I bet I could put some pep into him. Why I could-Well, sir, believe it or not maybe my prayer was answered, maybe it was mental telepathy or maybe the stars were just right. I do not know what, but anyway kind Providence delivered him into my hands. He slowly arose, came out into the yard where I was and sat down on a hive not more than six feet from where I was working and right in front of the hive I was about to open. He asked, "What are those things?" I could hardly realize anyone could be so ignorant as not to know what bees are but evidently he did not. Thinks I, "Well, old boy, right here is where you are going to get your first lesson in bee culture." I reached down and took hold of the bottom of the hive, raised it about six inches above the stand and let it fall back with a bang. Well, sir, it was a great success. No soldiers ever responded to the command "charge" more nobly than did those bees. It looked as though all the bees in that hive had been shot out of a cannon and their aim was perfect. They covered the face of that tramp till I could hardly see it! He gave a screech, fell over backward, tipping over the hive he had been sitting on. He made a sort of barrel roll, got on all fours and started through the sweet-corn patch with a speed that would do credit to a trained athlete. The bees in the hive he turned over joined in the frolic but these did not sting him in the face. The tramp vanished forever from my sight. After a time Mrs. Mechlin came out with a large tray filled with delicious food, hot biscuits, hot coffee and cream and a big slice of beefsteak. She asked, "where has my man gone?" I replied, "I do not know where he has gone but if he keeps on the way he started he is there before this no matter where it is." Then I added "Furthermore, Mrs. Mechlin, I am confident that for some time in the future our journeyman will take his hand-outs standing." To think she had prepared that excellent delicious breakfast for that worthless tramp. And it *was* a delicious breakfast. I ate it.

Linger

# The Professor and the Bees

One day a certain Dr. Hamilton called me up and informed me that he had just purchased some bees, five colonies in all, from a party living but two blocks from his home. He had moved the colonies home but reported that most of them had gone back to their former home and about a bushel of them were hanging on a post. He asked what he should do. I told him he had done too much already, but suggested that he move the colonies back, then after dark move them a couple of miles, then in a week or so bring them home. He thought that too much trouble and wanted to put a hive there for me to bring over a queen. This I did. The bees were furious and began stinging me before I got near them. I was slapping them right and left every time one lit on me sometimes before it stung but usually just after. Here the Doctor interrupted. "My friend, evidently you are not familiar with the psychology of the honey bee. The bee is an extremely intelligent insect and will never harm anyone unless it thinks you are going to harm it. Any quick motion you make causes the bee to believe you are going to harm it and it therefore stings in self-defense. Didn't you know that?" "No, I did not know that but I want to learn more about bees so wish you would teach me. Please take this smoker and drive the bees off from that post into the hive." Very well, I shall take pleasure in teaching you the fundamentals of bee psychology." He took the smoker and calmly walked toward the bees. Being sure there was going to be a scene I retired well out of range to watch developments and, let me say, things surely developed at a rapid pace. A bee stung him on the back of the neck but he went calmly on. Another stung him on the ear, still his serenity of disposition was undaunted. Then one gave him one of those double-strength hot ones right on the end of his nose. His placidity vanished instantly and he fairly exploded. He dropped the smoker and brought his hand to his nose with a vicious slap. In doing so he knocked off his sailor hat displaying a shiny bald head. The bees fairly covered his head and the professor went off on the run and took haven in a small house that had a half moon cut in the gable. Then I called to him, "Professor, don't you know you should never fight the bees? Their psychology will just not stand for it." As he rubbed the stings out of his head he replied, "But such action on their part is wholly incomprehensible and intolerable for I meant them no harm whatsoever. I only mean them good and wished to help them out of their difficulties and just see how unappreciative they have been in attacking me with no provocation whatsoever." "In other words, Professor, you no longer consider the bees the embodiment of superb intellectuality but just a diabolical pusillanimous palooka and the idea that the bee has any intelligence whatsoever is all hogwash is that right?" The Professor declined comment. He never went near those bees again. One of the neighbors said the Professor refused to have anything more to do with them as they were so unappreciative of his efforts in their behalf.

## Conclusion

Life to me has been such a paradox, so much happiness and joy yet so much misery and sorrow, that makes one wonder just what it is all about. I will close by quoting my favorite poem by Leigh Hunt, as it best gives my aim in life.

### Abou Ben Adhem

Abou Ben Adhem (may his tribe increase)
Awoke one night from a deep dream of peace,
And saw within the moonlight in his room,
An angel writing in a book of gold:-
Exceeding peace had made Ben Adhem bold,
And to the presence in the room he said,
"What writest thou?" The vision raised its head,
And with a look made of all sweet accord,
Answered, "The names of those who love the Lord."
"And is mine one?" said Ben. "Nay not so."
Replied the angel. Abou spoke more low,
But cheerful still; and said "I pray thee then,
Write me as one that loves his fellow-men."

The angel wrote and vanished. The next night it came again, with a great awakening light,
And showed the names who love of God had blessed,-
And, lo Ben Adhem's name led all the rest!

# *THE END*

# Addendum: The Better Queens Method

(condensed from Better Queens By Jay Smith)

### Transcriber's and Editors preface.

I found it difficult sometimes to figure out the next step from the entire book so I edited a condensed version of *Better Queens* because, while the full book is wonderful and educational and should be read before this condensed version, it chases a lot of "rabbits" and explains a lot of the reasons and explains a lot of the failures to help you avoid those. Assuming you have read the full text of *Better Queens*, and are convinced you want to follow his method and want it distilled down to just the method, here it is, still in Jay Smith's own words. When working through the method this makes a better "checklist". The pictures are available in the above text so refer to the full book for the pictures.

### The Breeder Hive

The cut (see above) shows the frame partly filled with wood, leaving a small comb five and one-half inches by nine and one half inches. These frames we shall call wood frames. It takes the queen but a short time to keep these combs filled with brood so she can devote her time to laying in our new comb. The standard size foundation will do very well but we prefer the size used in the Modified Dadant frames as it is deeper. A sheet is cut into four parts and fitted into standard frames. These frames of foundation may be drawn out in either the finishing colonies or in the front part of the breeding queen hive. When one of these new white combs is placed between the two combs partially filled with wood the queen at once begins to lay in it as the other two combs are filled with brood and the new comb is right between. This makes an inviting comb for the queen to lay in, which she does at once. A wooden partition separates the compartment containing the three combs from the rest of the hive. A cleat is nailed to the bottom of the hive below the partition. A space of an inch is left between the cleat at the bottom and the partition above. A zinc queen excluder is tacked over this opening. It is best to have the excluder at the bottom for if higher up the bees insist on filling the space with comb which they will not do when the excluder is placed near the bottom. When using the regular cover, the queen would sometimes get out of her partition due to the fact that wax or propolis would accumulate on top of the partition. To overcome this the partition is extended three-fourths of an inch above the edge of the hive. We use two covers, one for the back part and one for the front. Then the regular telescope cover is placed on top. With this arrangement the queen never gets out of her stall. The cut will show how the hive is constructed better than a description.

We use a jumbo hive cut down to fit the standard frames. One may use the standard hive but it will be necessary to nail cleats around the bottom to make room under the frames. The entrance is about midway in front of the hive.

### Keeping the Breeder Hive Stocked with Bees

If cells are started every day the bees in the breeder hive will rapidly decrease in numbers as the only emerging brood they will have is what comes from the combs in the wood frames, therefore, some way must be provided to keep up the strength of the colony. We open the starter hive in front of the breeder hive when we take the started cells out. Many young bees will take wing so when we remove the starter hive, these young bees will join the bees in the breeder hive. This is a great help but shortly they will need more bees in which case we shake the bees from not more than one comb from the starter hive. In this way their strength will be kept up. Sometimes in the past we dumped the bees from the starter hive in front of the breeder hive. This proved unsatisfactory as so many bees were apt to kill the breeder. When they did not the breeder was apt to stop laying for a time. Adding the smaller number of bees has never given us any trouble.

### The Feeder

This is an important part of the breeder hive. When no nectar is coming in the bees must be fed if we are to get that lavishly fed larvae we have pointed out is so necessary for best results. It is built on the back of the hive. A hole is bored in the back of the hive so the bees can get into the feeder. Slats are put in as illustrated, then melted wax is poured in to make the feeder waterproof and also to wax the slats tight to the bottom in order that no syrup or honey can get under these slats to sour. A cleat is nailed to the hive just above the feeder to keep out the rain. When feeding, the cover of the feeder is slid back and the honey or syrup poured in. As there is no bee disease within many miles of our apiary we feed off grades of honey. Honey is better food for the bees than sugar syrup but if disease is present sugar syrup should be fed. However, from years of observation I am confident the nurse bees cannot secrete milk with proper nourishment when sugar syrup is fed. I was surprised when I first put honey in the feeder. It was during a robbing season and I wanted to test it for I had a theory feeding with it would not cause robbing. It proved better than I expected for by looking at the bees at the entrance you could not tell they were being fed. With any other feeder I ever tried we had to feed at night or the robbers would rob out the colony. What causes robbing is that the bees get daubed with honey and then go out in front to get more air. The bees from the other hives pounce upon them and then proceed to clean up the whole hive. With our feeder few if any bees get daubed and those that do have to pass through a cluster of bees, then through the queen excluder and then another cluster of bees. By that time they are licked clean and do not have to go out for an airing.

### Stocking the Breeder Hive

In case your breeding queen is in a fairly strong colony, the bees and brood are placed in the front compartment and one frame of brood and the queen is put in the back compartment. The two frames fitted with wood are placed in the back compartment with the frame of brood between them. Feed is then placed in the feeder. As soon as the queen has eggs and brood in the two outside combs, the frame of brood is removed and a new white comb is placed between the two combs. In case the colony is strong, most of the brood may remain in the front compartment, but later most of this must be removed to make room for the combs containing eggs. As stated, we should have several drawn combs ahead. When the brood in the front compartment is removed frames with sheets of foundation may be given. As the bees are being fed, they draw the foundation readily. The comb given should be left 24 hours when it will be well filled with eggs. It is then placed in the front compartment and another frame of new comb is given. This is repeated every day until there are three frames containing eggs in the front compartment. When the fourth comb of eggs is put in the front, the first one given will now contain tiny larvae just out of the egg and as this is practically all the brood that large numbers of bees have, the tiny larvae will be found to be fairly flooded with bee milk. In giving comb to the breeding queen, it should be trimmed back to the foundation provided the bees have built comb at the edges and bottom of the foundation. The queen prefers to lay in natural built combs rather than those built from foundation. I have used all kinds of foundation and find this to be true. The comb shown was not trimmed back so it can be seen that the queen laid in a sort of horseshoe shape.

The bees do not build cells with such regularity as the cells in foundation, therefore such comb is difficult to cut as many larvae will be destroyed. When the eggs are laid in cells drawn from foundation it is much better if the row of cells is uniform and easily cut in a straight line. Now let us examine the larvae in the cells to see if they are getting lavish feeding, the necessity of which we have so frequently stressed. The larvae are so small they can hardly be seen with the naked eye. They are flooded with bee milk fully as much as larvae in swarming cells. In these same cells they will remain while the bees go right on feeding them. Let us now prepare to start those larvae on the road to queendom.

### The starter hive

Many years ago Henry Alley used what he termed "the swarming box". He did not start cells in it but confined bees in it for some hours and at night dumped them into a hive with no brood or queen. He seemed to think it necessary to keep the bees queenless for some time to get them to accept the cells. Then the prepared cells were given to the bees in a hive containing two frames of honey and pollen. Mr. G.M. Doolittle said the bees would accept cells better if kept queenless for three days. In my experience I find the bees accept the cells just as readily after an hour or two as they do if left queenless for a longer period. We usually keep them confined in the starter hive from one to two hours which gives them plenty of time to clean up any honey that may have been spilled on them when shaking them into the starter hive. In later years Eugene Pratt gave the cells direct to the bees confined in a box and he called it "The swarm box". We prefer to call it "The Starter Hive" as it is used for starting cells and has no relation to a swarm. Our starter hive is made wide enough to hold five standard frames. The total depth is 17 inches. It has a wooden bottom with cleats at the sides making it suitable for feeding or watering bees. Three frames of honey and pollen are put in *but no brood*. In case no nectar is coming in, the bees that are to be used in stocking the starter hive should be fed three days before using them and this feeding should be continued each day until they are shaken into the starter hive.

### Food and Water Necessary to the Bees in the Starter Hive

A matter that I believe most of us have overlooked is the necessity of food and water for the confined bees. All have observed how bees carry water in abundance when rearing brood. Now this water is even more necessary when rearing queens for the bees require more water when producing the milk in abundance. The best way to provide this is to fill the center comb with diluted honey. This not only provides the necessary water but the food it contains acts as a flow of nectar. In case some 50 cells are to be started from five to six pounds of bees should be used. In case from 75 to 100 cells are to be started from 8 to 10 pounds of bees should be put in. If larger number of bees are put in the starter hive should be kept in the shade or some cool place. The screens on the sides provide an abundance of ventilation.

### Getting the Bees with Which to Stock the Starter Hive

The colonies are all in two stories with queen excluders between the two hives. The queens are confined below and the top story is kept well filled with brood. All that is necessary to do in filling the starter hive is simply to shake in the bees from the top story. The bees will be depleted somewhat, as some are used to keep up the strength of the breeder hive and the finishers, so unsealed brood should be given to these colonies from other colonies. We use the bees from each colony once in every four days. This is less work than it would be to use a larger number.

### Preparing the Cells

Let us now consider the larvae in the breeder hive is ready for use. First some wax should be melted. An electric hot plate or a sterno stove with canned heat may be used.

The comb from the breeder hive is now cut into strips one cell wide. With a small paint brush, wax is painted on the cell bar and the strips of cells placed on the waxed bar. Then with the brush wax is painted on the sides of the cell strips. The cells are then cooled in the wooden tank and more wax painted on till both sides of the strip of cells have enough wax to make the cells strong enough so they will not mash down when inserting the spike. Wax should be cool for painting the cells, nearly ready to solidify. We use a wooden tank four inches wide inside measurement, and half an inch longer than the cell bars. A cleat is tacked in one end under which the end of the cell bar is placed. At the other end of the tank is a nail driven through from the side extending about two inches into the box. As soon as the cells are waxed the bar is shoved under the wooden cleat at one end of the tank and the other end is placed under the nail at the other end. Then cold water is poured in the tank till it covers the melted wax. It is well to put ice cubes in the water for we want the cells to be firm so the larvae in the cells can be destroyed without mashing down the comb. While we prepare the cells as quickly as possible in order that they may be in care of the bees without delay, it has been astonishing to note what gluttons for punishment these well-fed larvae are. Frequently we have more cells that we have use for so instead of using the larvae, we set the comb aside for the larvae to die in order that we may use the comb again. Often after four days, when we give the comb to the bees in the breeder hive, some of the larvae will be rejuvenated or resuscitated, or whatever you wish to call it. An interesting feature is that although the larvae were away from the bees four days, they had not increased in size a single bit. This shows that larvae away from bees do not grow, so we should keep them away from the bees as short a time as possible.

It will be necessary to destroy some of the larvae in the cells in order to make room for the queen cells. We have tried many methods but find the best way to destroy the larvae is to insert a spike in the cells and punch out the bottoms. An ice pick may be used if the point is broken off. We usually destroy two cells and leave one which gives plenty of room for the fine large queen cells. Sometimes when we are short of cells we destroy every other one. Some will be built close together and will be destroyed in cutting but one can get more cells in this way.

### Putting the Prepared Cells Into the Starter Hive

We use two bars in each frame, the space above and below the bars being filled out with wood. The starter hive is given a little jar on the ground so the bees will fall to the bottom and not fly out. The two frames of cells are set in place, the comb containing thin honey being in the middle. Cells will be built better in the middle of the combs as shown, for where the cells are near the top of the frame the bees are apt to build comb over them.

### Program of These Rejoicing Bees

Now what is the first thing the bees do upon receiving these prepared cells? They first unload the accumulated milk in their glands by feeding the larvae. Next, *they begin tearing down the walls of the worker cells.* It has been stated by many that the bees enlarge the worker cells. This they can not do. They tear down the worker cells and build queen cells over the larvae. In this way they build the cells as they want them and do not have to put up with any man-made contraptions.

When starting cells from worker cells the bees first "raze" them and then "raise" them. This reminds me of an incident that happened years ago. I told my wife that a virgin had gotten into the hive where I had a lot of queen cells and this virgin was razing cells. She thought I said the virgin was "raising hell" so I let it go at that for I rather liked her version of the incident.

### Finishing the Cells

While the bees are now busily engaged in remodeling the cells, we must prepare to have them finished, for they should not remain in the starter hive more than 24 hours for the best results. Again I shall tell what we do and then tell why we do it. The best colony for finishing cells *is a strong queenless colony.*

The queenless colony that is to finish the cells must contain at least five pounds of bees and several combs of mostly capped brood. If no nectar is coming in, this colony must be fed liberally. Again, to be extra conservative, we have two colonies to finish the cells that have been started. If this is properly done you can produce cells every bit as good as those built by the bees during swarming. Bees must be added to keep up their strength. It is some help to them by giving them the cells with the bees that are with the cells. When more bees are needed the starter hive is dumped in front of the finishing hive.

We use a single story to finish the cells unless there is a heavy honey flow on, in which case we add a second hive body but have the cells finished in the bottom hive. If we start four bars of cells in the starter hive it is best to give them to two finishing colonies. As the bees finish a job that has been started we must be very particular to see to it that they are in the best possible condition to finish the cells properly. In case we wish to start cells every day we have three sets of finishers, two hives in each set. In case one is rearing queens on a very extensive scale he can have four hives or more in each group and use two or more starter hives. After the cells have been in the starter hive overnight we lift out the frames of cells and give one to each of the finishers. The adhering bees are given to the finishers which helps keep up their strength. The frame of brood and cells are arranged as follows: At the side of the hive nearest to you is a frame of capped brood, then a frame of cells. Next day, group number two are given bars of cells, then next day group number three are given two bars of cells. Now, when you start the fourth batch of cells, the ones in the first group are sealed or nearly so and are given bars of cells, a frame of capped brood separating them from the first frames given. This is repeated till all finishers have three frames of cells each. When ready to give the next frames of cells, those first given now are what is known as "ripe cells." They must be taken out to make room for the next frames, and are ready to be introduced to mating hives. These cells will hatch the following day.

The photograph on page 60 shows the arrangement of frames of cells as they are painted white that they may be easily seen. From this on, when a frame of ripe cells is taken out another frame with started cells is given and this may be continued throughout the season. As the bees in the finishers become weak more bees must be added from the starter hive and capped brood given from other colonies. In case you do not need so many cells you can skip a day and the space in the finisher is filled by moving the combs over and should there be a honey flow an extra comb should be given to prevent the bees from building combs in the space.

### Caution

Three or four days after giving brood you should look over the frames to remove any cells that may have been started. It is astonishing to note how many accidents can happen in those finishers no matter how careful one may be. We tack a zinc queen excluder to the bottom of the finishers to keep out virgins; for if there is any place one can get in is sure to do so. By making the hives bee-tight and by examining the brood carefully you will have no accidents, or rather I should say, few.

### The Adolescent Mating Hive

The most successful small hive I ever used is what I term my adolescent hive. It is larger than the baby hive. It is made to hold three frames 5 x 8 inches. It is made of full three-fourths inch material and the cover is made of 2-inch material.

A small mating hive should not have a ventilator, as the air will blow in through it and cause the bees to realize their weakness and they will abscond. For an entrance a half-inch hole is made in the center of the hive. This hole is slanted upward to keep out the rain. If the entrance is placed at the bottom, the ants or robber bees will kill out the bees if they get weak.

When feeding, a battery filler can be used and a little food may be squirted into the entrance at night. A package box which is used for bringing bees from the outyard is made with wire screen sides as shown in the cut. Half a pound of bees will nicely stock this hive. A little feed is given, a ripe cell is placed between the combs, and the package box is placed on top. The entrance should be kept closed for 24 hours if the bees are taken from the outyard. If taken from the home yard the bees should be confined 48 hours, or many will return to their former colonies.

After confining them for the required time, the entrance is opened and the package box removed. To those who prefer a small hive this is by far the best of anything I ever tried. Even this hive requires close attention to see that it is kept in the right condition. We have wintered queens in such hives here in Florida but the results were rather uncertain so we have adopted a larger mating hive, one holding five standard frames. These hives are placed on concrete standards, two hives on each standard. The covers are made of one-and-one-half inch material. This cover keeps them warm in the winter and cool in the summer. They are heavy enough so no wind will blow them off. At first we formed them in spring. This weakened the colonies in the outyard to such an extent that they gathered no honey. Now we form the mating hives in the fall after the honey flow is over. We proceed as follows: As the bees in the outyard must be requeened, we take a package of one and one-half pounds of bees and the old queen. The bees and queen are put into the mating hive. As they have a queen it is not necessary to confine them if they are put in after dark. A young queen is introduced into the colony from which the package was taken. With a young queen, the colony soon makes up for the loss of the bees. After the old queen in the mating hive has been laying for a week or so she is removed and a ripe cell given. With this management, the colonies in the outyard are actually benefited. As we winter the colonies in the mating hives there will not be many to establish in the fall, only the ones that have run down for one reason or another. Some mating hive colonies may be too weak for wintering. In such cases a package with queen is put in the same as though the hive was empty. When there is no honey flow, our five-frame hive requires more feed than the smaller one. On the other hand the bees in our larger hive will gather more honey and therefore need less feed. While rearing queens we have but four frames in the hive as it is much more convenient to handle the frames. After queen rearing is over in the fall, an extra frame of honey is put in. A few may die out in the winter for one reason or another, but the rest will be so strong that they may be divided to make up the loss. When introducing cells, when no honey is coming in, we feed the night before giving the cells. Should any be short of stores, they are given an extra amount of feed. The entrance to this hive is in the center of the front. It is one and one half inches in diameter and the hole is slanted upward to keep out the rain.

### Introducing the Cells

In introducing cells no protectors are used but we simply press the cells lightly against the combs, placing them near brood as the bees will care for them better.

### Stocking the Mating Hives

In stocking the mating hives it is best to take them from an outyard as they do not have to be confined so long. We do not weigh the bees but measure them instead. We have what we call our bee hopper. It is four and one half inches wide. The bottom is covered with galvanized iron. A scoop is made just wide enough to reach across the inside of the hopper. For the five frame hives we use about a pound and a half of bees.

### How Long Should the Bees be Confined to the Mating Hive?

In case bees are taken from the outyard they should be confined to the mating hive one day and released after dark the day following. In case bees are taken from hives in the home yard close to where they are to be established they should be confined two days and released after dark. Most of them will stay but a few will be gifted with remarkable memory and will go back, but there will not be enough to materially weaken the bees in the mating hives.

## Mating Hive Records

One must have some system of knowing what the conditions are within the mating hives. We have used a number of systems and out of all of them we have worked out a system that gives such perfect satisfaction that we never expect to make a change. We need to know the conditions within the mating hive for a period of two weeks, for by that time the virgin should have been mated and by that time she is a laying queen. Therefore, we have a row of copper tacks across one side of the hive, 14 in number, one for each day of the week for two weeks. These tacks are driven about halfway in. The first tack at the left represents Monday. By putting a mark over Thursday's and Sunday's tacks, we can see at a glance what days we are working on. Tags are made of galvanized iron two inches square. A square hole is made in the center of the tag. The corners are bent forward a little for easy handling. They are first dipped in yellow quick-drying enamel. After drying, each corner is dipped in enamel of the following colors. First black, then red then white, then blue reading to the left. Now let us consider the hive and it's signal relation to the queen. Let us say we introduced the first cell on Monday. The tag is placed on the tack at the extreme left and turned to red, indicating a cell has been given. We examine the hive Wednesday and if the virgin is there the tag is turned to white. Should the virgin not be there, the tag is turned back to black indicating the hive is queenless. When another cell is given, after the first one is not accepted, it is put in on Wednesday so the tag is moved to the third position and turned to red. This process is continued throughout the season. A picture of our mating hive has already been given. To indicate the condition within the hive we have blocks four inches long painted black so they may be seen at a distance. A number of these blocks are distributed among the hives, being placed on the stands between the hives. In case the hive can spare honey the block is placed lengthwise of the hive at the extreme left. If it needs honey it is placed crosswise in the same position. The block shown in the cut shows the hive can spare both brood and honey. If placed crosswise at the extreme right it indicates that the hive needs brood. If placed lengthwise on the same position it indicates the hive can spare brood.

### Bee Candy

Candy for the introducing cage or for mailing cages is made by mixing XXXX powdered sugar with honey. This should be kneaded as stiff as possible. If the honey is heated it will absorb more sugar. If one has bee disease, or is unable to get a health certificate before shipping, candy may be made of invert sugar and such candy is nearly as good as that made with honey. To make invert sugar, put half a pound of water in a kettle and let it come to a boil. Then add one pound of granulated sugar and stir in one-fourth teaspoon of tartaric acid and boil till a temperature of 248 degrees is reached. A candy thermometer is necessary for this. No certificate of health is required by the postal authorities when using invert sugar candy. All that is necessary is a statement that there is no honey in the cage. After the invert sugar is cooled it is used the same as honey in making bee candy.

### Shipping Queens

Our experience has taught us that the reason so many queens die in the mail is lack of ventilation or because they are overheated. We have found that by wrapping the cages in cloth, they go through in good condition if they are not too long on the road. We also ship in mailing tubes which give even better ventilation. Shipments made with mailing tubes have gone to 43 foreign countries by air with almost no loss.

### A Review

As we thought it necessary to do a lot of explaining as we went along, we will now give the procedure step by step:

1. Stock the breeder hive as described, putting queen and one frame of brood and the two wood-comb frames in the back compartment, putting the frame of brood in the center.

2. Prepare as many colonies as are needed by putting them in two story hives with brood above the queen excluder and the queen below.

3. Cut Modified Dadant foundation in four pieces and put them in frames and get this foundation drawn in any colony suitable.

4. Take out the frame of brood from the back compartment of the breeder hive and put it on the other side of the partition, being sure the queen is in the back compartment. Then put in a frame of the drawn foundation. Leave it there for 24 hours, then put it on the other side of the partition. In case you wish to start cells every day, put in another frame of drawn foundation in the back compartment. Repeat this every day until the

fourth comb is placed the other side of the partition, when the first is ready for use.

5. Put three frames of honey and pollen in the starter hive and put some thin syrup or diluted honey in the middle comb. Then shake in bees in the amount of five or six pounds from the hives previously prepared. Leave them three or four hours.

6. Take out frame of young larvae and cut into strips, mount them on bars and give them to the bees in the starter hive.

7. Next day remove the queen from two colonies to be used as cell finishers and give to each colony one frame of cells and dump the starter hive in front of the colonies from which the bees were taken.

8. When all finishing colonies have three frames of cells each, and the fourth is ready, remove the first given and introduce the cells to mating hives.

## Starting Cells Every Third Day

In starting cells every third day, but two finishing colonies will be needed. In this case you put in a new comb in the breeder hive every third day and when it is placed the other side of the partition, an old comb or a frame of brood is placed in the back compartment between the two wood-combs. There are both advantages and disadvantages in starting cells every third day instead of every day. We are running more to the third-day plan. One advantage it has is that the drawn sheet of foundation may be left for the queen to lay in longer. Another is that as cells are introduced every third day, in introducing the cells, the ones introduced three days before can be examined. If using the third-day system, in order to keep the record straight, one should write down on a piece of paper the letters C S I, C S I in rotation, these letters being placed at the left hand of the paper, one below the other down the paper. C stands for comb and indicates that you gave a new comb to the queen in the breeder hive. S means that you start cells, and I means you introduce cells. Put the date and the day of the week after each letter. Of course in starting there will be no cells to start or introduce, but after some time the letters will have a meaning. In case one started cells regularly every third day there would be little use for the above memorandum as the condition of the cells in the finishing colonies would indicate what is to come next. However, in case you slipped one starting, all cells would look much alike and you might get things mixed up. At any rate I have done so and when I came to giving the started cells to the finishers I found the hive full of virgins!

## Modifications for the Small Beekeeper

For the commercial queen breeder, or for the honey producer who wishes to rear 200 queens or more I strongly recommend the management and equipment that have been described. For the small beekeeper, or the back-lotter who wishes to rear a dozen or maybe 50 queens in a year, no special equipment will be necessary. Under such circumstances it is best to rear queens during a honey flow as conditions are more favorable and queen rearing is a simple matter compared to rearing them when there is no flow. When the honey flow is just beginning is the best time to start as there is an abundance of pollen coming in along with the nectar, and pollen is just as necessary as nectar.

For starting cells, first get the foundation drawn as described. Next set to the rear the hive containing your breeder, with the entrance in the opposite direction from the original position. In its place put another hive and in it put two frames solid with honey and pollen. Shake in half of the bees from the breeder hive and put the breeding queen into the prepared hive. Within 24 hours she will have a good patch of eggs in the new comb. The comb should be placed over an excluder on a strong colony. In case you do not wish to start more cells immediately, this hive with the two combs and the one with the eggs may be set over the hive formerly containing the breeder, setting this hive on its former stand. The two hives must be separated with an excluder, with the queen in the bottom hive. This colony must be well fed for we want the larvae that will soon come from the eggs to be well fed. As soon as the eggs hatch, turn a strong colony to the rear and in its place set a hive into which you put two frames of honey and pollen then shake out half of the bees from the hive you have moved into the hive with the two combs. In case the colony is not of sufficient strength, two colonies may be set to the rear and half of the bees from both colonies shaken in. Next cut the comb into strips as has been described and give a frame of two bars of prepared cells for this colony to both start and finish. In case one prefers to use a starter hive, one can be made by merely tacking a screen to the bottom of a hive. If this is done the hive should be set on blocks high enough to afford plenty of ventilation.

## Mating Hives for the Small Beekeeper

In case increase is desired a hive may be moved to a new location and another hive set in its place. Into this hive put two or three frames of brood with adhering bees from the hive moved, being careful not to get the queen. It is well to wait two or three days after dividing this colony before giving a cell for bees will keep returning from the hive moved and are apt to destroy the cell given. As soon as the queen has mated and laying, the hive should be filled out with combs or frames of foundation or starters. In case feeding is necessary a good way to feed is to tip the hive back a little and under the front place a block. Then at night pour in a little syrup. Be careful not to put in more feed than the bees will consume during the night for if any syrup is left it will incite robbers as a bottom board feeder is the worst of any feeder to cause robbing.

## Mating Queens in an Upper Story

In case one does not wish to make increase, a good way to mate queens is in an upper story of a two-story hive. A frame covered with wire screen should separate the two hives. Always use a wire screen and *never a queen excluder.* Some in the past have advocated using a queen excluder but in the great majority of cases the bees will kill the virgin as soon as she emerges. In the frame of the wire screen an opening of half an inch should be made for a flight hole. This entrance should be to the rear of the hive below or the queen in returning from her mating flight is apt to get into the hive below.

The entrance to the top hive should be closed for two days, otherwise too many bees will join the colony below. Syrup or honey should be put into one of the combs and a ripe cell given. After two days the entrance should be opened and the virgin will soon fly out and mate. When the honey flow permits, this is an excellent way to requeen the colony for we have *two laying queens* instead of having it queenless while introducing a queen or cell. The bees in the top hive will not suffocate from having the entrance closed for the screen will afford abundant ventilation. The best way to put syrup into the combs is with the use of a pepper box feeder. That is a mason jar with the cap full of holes. With this the syrup may be shaken into the combs. If you merely want to requeen, all that is necessary is to hunt up the old queen below and remove her and take out the screen. The bees will not fight as they have kept up their acquaintance through the screen. In case increase is desired, just move the bottom hive to a new location and set the top hive on the old stand. The lower hive should be moved as it is much stronger than the top hive. The bees for a time will try to get into the back of the hive but after a time they will make a chain around to the front door and all will be well.

## Requeening by Giving a Ripe Cell

As a rule we do not recommend giving a ripe cell to a full colony, as it weakens the colony by being deprived of a laying queen for nearly two weeks. Again, a strong colony does not accept a cell as readily as a weaker one, and if the cell is not accepted it will further weaken the colony. However, when the honey flow is over and more bees are not needed immediately, it is often desirable to requeen by giving a ripe cell. As it is difficult to find a virgin in a strong colony, one may give a cell and without looking for the virgin another cell is given three days later. In a week after the first cell is given, the colony may be examined and if there are no queen cells started, you may know a virgin is present. In case both cells are destroyed, or should the virgin be lost in mating, it is best to introduce a laying queen as the colony would become too weak if given another cell.

## About the Author

Jay Smith was an active writer in the bee journals of his day and wrote two of the most loved and used of the queen rearing books still being used and followed by many today. He is most famous for his other book, Queen Rearing Simplified, but this book is the culmination of his work in queen rearing.

www.ingramcontent.com/pod-product-compliance
Lightning Source LLC
Chambersburg PA
CBHW031521270326
41930CB00006B/473